CW00602949

1996
First
Bus
Handbook

October 1995

British Bus Publishing

1996 FirstBus Bus Handbook

The 1996 FirstBus Bus Handbook is a special edition of the Bus Handbook series which contains the various fleets of FirstBus plc. The Bus Handbook series is published by British Bus Publishing, an independent publisher of quality books for the industry and bus enthusiasts. Further information on these may be obtained from the address below.

Although this book has been produced with the encouragement of, and in co-operation with, FirstBus plc management, it is not an official FirstBus fleet list and the vehicles included are subject to variation, particularly as the vehicle investment programme continues. Some vehicles listed are no longer in regular use on services but are retained for special purposes also out of use vehicles awaiting disposal are not all listed. The services operated and the allocation of vehicles to subsidiary companies are subject to variation at any time, although accurate at the time of going to print. Livery details given aim to show standard fleet colours and significant variations. Minor variations or local identity and marketing schemes are not all included and in some cases older liveries being phased out are not shown. The contents are correct to October 1995. The major restructuring of Yorkshire Rider Limited into new operating divisions was ongoing while this book was being prepared.

To keep the fleet information up to date we recommend the Ian Allan publication, Buses published monthly, or for more detailed information, the PSV Circle monthly news sheets.

Principal Editors: David Donati and Bill Potter

Acknowledgements:
We are grateful to Cliff Beeton, Mark Jameson, Andrew Jarosz, Ken Jubb,
Tony Wilson, the PSV Circle and the management and officials of FirstBus
and their operating companies, for their kind assistance and co-operation in the compilation
of this book.

The front cover photo is by Tony Wilson
The rear cover photographs are by Richard Godfrey and David Cole

ISBN 1 897990 27 8
Published by British Bus Publishing Ltd
The Vyne, 16 St Margarets Drive, Wellington,
Telford, Shropshire, TF1 3PH
© British Bus Publishing Ltd, October 1995

FIRSTBUS

FirstBus plc is the second largest UK bus company and came into existence in June 1995 as a result of the merger of Badgerline Group plc and GRT Bus Group PLC. The geographical coverage of the company runs from Aberdeen in the north to Cornwall in the south, and from Wales to Norfolk with an operational fleet of some 6000 vehicles.

FirstBus is a Scottish registered company and has its corporate headquarters at Badger Manor near Weston-super-Mare where the Chairman, Trevor Smallwood OBE, and the Finance Director, Tony Osbaldiston, are based together with FirstBus finance and administration functions. Operational Headquarters are at Aberdeen where the Deputy Chairman and Chief Executive, Moir Lockhead, and Group Commercial Director, Robbie Duncan, are based.

The combined fleets have been organised into three divisions. The Scottish operations, North of England - PMT Group and Rider Group - and South of England and Wales.

Since June the operations of District Bus of Chelmsford and People's Provincial have been added, the latter following an agreed take-over which took place while this edition was at the printers. The Frontline operation based close to Lichfield was sold to British Bus plc.

Two innovative and complementary guided bus schemes have been developed, one in Leeds, the first phase of which is operated by Yorkshire Rider using new Scania single deck buses and the other in Ipswich marketed as SuperRoute 66 which is operated by Eastern Counties. A further plan is the proposed use of a single-track railway line between Portbury, in Avon, and Bristol, which would provide a guided busway into the city centre.

Both GRT and Badgerline have invested in new vehicles, choosing different suppliers during the last year. Badgerline have taken quantities of Dennis/Plaxton buses with both the Dart and Lance delivered in large numbers. GRT took the Mercedes-Benz O405 with bodywork from Optare and Wrights, Scania L113s and ordered the first of the Volvo B10Ls with Alexander Ultra bodywork, which, like several GRT orders, feature double glazing, air conditioning and continental seating.

At the time of the merger GRT comprised the operations of Grampian Transport, Northampton Transport, Leicester CityBus which are primarily city based while Midland Bluebird, Lowland, Eastern Counties and SMT provide town, inter-urban and rural services.

GRT Bus Group PLC came into being on 23rd January 1989 following a management and employee buy-out of Grampian Transport. However, the origins of the Group can be traced back to 1898 when the assets of the Aberdeen city tram services were transferred to the town council.

The Chairman of FirstBus is Trevor Smallwood OBE *(left)* and Deputy Chairman and Chief Executive, Moir Lockhead are seen on a visit to one of the operations. The geographical coverage of the company runs from Aberdeen in the north to Cornwall in the south, and from Wales to Norfolk with an operational fleet of some 6000 vehicles.

Following deregulation of the bus industry in 1986, Grampian Regional Transport was incorporated as a subsidiary of Grampian Regional Council. In December 1987 Grampian Transport acquired the business of Mairs, a company primarily involved in private hire coach work.

Further expansion followed in July 1989 with the acquisition of Kirkpatrick of Deeside, a company founded in the 1950s by Tom Kirkpatrick which provided an addition to the coaching activities. Coach services were introduced later in 1989 with vehicles from Mairs fleet liveried for Scottish Citylink with daily duties taking the vehicles to Birmingham, Middlesborough and Glasgow.

In June 1990 it was announced that GRT's bid for Midland Bluebird had placed it as the preferred bidder. This was just one of the operators within the Scottish Transport Group being sold by the Government. The deal was completed in September and doubled the number of vehicles.

The Group underwent further expansion with the acquisition of Northampton Transport from Northampton Borough Council in October 1993 and 93.6 per cent of Leicester CityBus from Leicester City Council. The balance of the ordinary share capital was retained by Trent Motor Traction Company Limited. Both Northampton Transport and Leicester CityBus were incorporated under the 1985 Act and have been in the passenger transport business since the turn of the century.

In 1994 the directors believed that there remained considerable opportunities to acquire both publicly and privately owned bus operators and took the decision to float GRT Bus Group PLC on the stock exchange. The placing also provided an opportunity for the existing shareholders to realise part of their investment, though the directors and employees continued to hold a significant proportion of the equity. The successful flotation raised £17 million of new money and dealings commenced in GRT's shares on 5th May 1994.

The first major acquisition following the flotation was Norwich based Eastern Counties with a fleet of 375 buses. The acquisition of SMT, which operates public services in Edinburgh and the Lothians was completed in October 1994. At the time of the acquisition SMT had a fleet of approximately 370 buses. Following quickly on from the acquisition of SMT was the announcement of an offer for Reiver Ventures Limited, trading as Lowland Omnibuses Limited which operate bus services in the Borders region. Reiver Ventures Limited was formed to effect the management and employee buy-out of Lowland Omnibuses Limited as part of the privatisation of the bus operating subsidiaries of the Scottish Transport Group. The acquisition of Lowland which operated 160 buses by GRT was completed in November 1994.

Badgerline started operations in April 1985 when the country operations of Bristol Omnibus Company Limited, then trading as Bristol Country Bus, were re-launched and re-branded as Badgerline. A striking new livery of yellow and green, (also white on minibuses), was introduced, together with the now familiar Badger motif.

Badgerline Limited was formed as a separate company later that year within National Bus Company and commenced as a separate entity from Bristol Omnibus from 1st January 1986. The company opened its Head Office at the aptly named Badger House in Oldmixon Crescent, Weston-super-Mare. It inherited depots at Bath, Bristol, Weston-super-Mare and Wells.

Under the Government's privatisation programme for the industry, Badgerline became the second bus company to be privatised when on 23rd September 1986, it was acquired from National Bus Company under a management and employee buy-out. Eighty staff participated in the buyout by purchasing shares in the company. At that time Badgerline operated 400 buses, employed 950 staff and had a turnover of £15 million.

From those early beginnings, the Group grew into one of the largest in the UK bus and coach industry, principally by the acquisition of other bus companies.

In preparation for this expansion, the holding company, then known as Badgerline Holdings Limited, was separated from the original

Badgerline operations and moved to the present Group Head Office at Badger Manor, Edingworth, near Weston-super-Mare in 1987.

The first expansion of Badgerline took place in 1987 when 39 per cent interest in Western National was acquired from NBC with the balance held by Plympton Coachlines and financial institutions. The holding rose to 67 per cent in 1988, and the balance was acquired in 1989. Western National was one of four parts formed from the original Western National Omnibus Company in 1983 and was responsible for the operations in south Devon and Cornwall, including a share in the Plymouth joint services. The later arrangements were suspended with the introduction of deregulation in 1986. Local operators Grenville of Camborne and Robert's Coaches of Plympton were added to the company in 1988, the names continuing to be used on vehicles allocated to certain duties.

In 1988, the group acquired Midland Red West Holdings Limited, the holding company of Midland Red West and City Line. Midland Red West was the first of the Midland Red companies to be sold as part of the NBC disposals, secured by a management/employee bid in December 1986. Included with the company were all coaches that ran under the Midland Red Coaches name.

The principal operating area was Hereford and Worcester, though a base in Birmingham allowed the company to expand into the West Midlands metropolitan county, where a considerable fleet now operates on commercial and tendered services. Midland Red West took the rump of the Bristol Omnibus Company in 1987 in a joint bid with its management. These operations concentrated on City services in Bristol and the fleetname City Line was adopted. With the merger of Badgerline and Midland Red West Holdings in 1988, both city and country services in Bristol were once again brought under the same management.

In early 1990 the group extended its geographical reach with the acquisition of South Wales Transport and the associated fleets of Brewers and United Welsh Coaches. SWT was privatised through a management buy-out in 1987 since when it took over the local operations of Brewers of Maesteg. Some two years later a reorganisation of the Port Talbot operations moved vehicles and crews from SWT to the Brewers subsidiary. Brewers also commenced a number of routes previously operated by the now defunct National Welsh company in the Bridgend area. The reorganisation also transferred United Welsh Coaches to Brewers while still retaining their trading identity. SWT now concentrates on the Swansea area and south-west Wales where much of the local operation is based on minibuses and Darts.

The next operator to join the group in April 1990, was Eastern National, which operated in Essex and parts of Hertfordshire. Eastern

In early 1990 the group extended its geographical reach with the acquisition of South Wales Transport and the associated fleets of Brewers and United Welsh Coaches. SWT was privatised through a management buy-out in 1987 since when it took over the local operations of Brewers of Maesteg. Brewers 598, UAR598W is seen at Neath. *John Jones*

National was purchased by its management on Christmas Eve 1986. Expansion into London was achieved following successful tendering under the London Regional Transport tendering arrangements. To meet this expansion two depots were opened in London, at Walthamstow and Ponders End. Some three months into Badgerline ownership Eastern National was split into two companies. Thamesway was set up to cover the London area operations and the southern part of Essex while Eastern National retained operations in north Essex and parts of Hertfordshire. A livery of crimson and yellow was applied to the vehicles transferred to Thamesway.

June 1990, and Wessex of Bristol joined the fold with a fleet of 33 coaches used mainly on National Express contracts. Wessex also has its heritage as part of the original Bristol fleet and this operation helped expand the Badgerline role within the National Express service network. Subsequently, Wessex has developed into local bus operations in Bristol, branded as Sky Blue.

In November 1993 Badgerline Group was floated on the London Stock Exchange. The reasons for this move was given as a desire to make further acquisitions and to reduce the high level of gearing resulting from the earlier acquisitions. Some £36 million of new capital was raised which would allow the Group to finance future investment and acquisi-

Latest arrivals with Eastern National are nineteen Volvo B10Ms with Plaxton Premier bodywork. Seen on its first day in service is 609 N609APU. These are liveried in yellow and orange City Saver colours for the Southend service. *Colin Lloyd*

tions. At this time the group fleet consisted of 2359 vehicles; 395(17%) double-deck, 831(35%) single-deck buses and coaches, 1006(43%) midibuses and 127(5%) minibuses.

On 22nd February 1994 it was announced that Badgerline proposed the acquisition of the PMT group, the supplier of bus services in the north Staffordshire and south Cheshire region. The PMT group was formed in 1986, again to enable the management to acquire the company from the former National Bus Company. PMT had grown since then with the acquisition of parts of the Crosville business operating on the Wirral and Chester, and later acquisition of Pennine Blue in Ashton-under-Lyne. Included in the deal were the now dormant names of Crosville Ltd, S Turner & Sons Ltd, Paramount Leisure Ltd, plus the repair and maintenance company PMT Engineering Limited.

The handover of PMT was completed on 10th March and less than two weeks later it was announced that Badgerline had offered to acquire the whole of Rider Holdings Limited, through a new subsidiary, Badgerline Yorkshire. The fleets involved were Yorkshire Rider, Rider York and Quickstep, though the Yorkshire Rider fleet has subsequently been split into four all under the Rider Group name.

Two innovative and complementary guided bus schemes have been developed, one in Leeds, the first phase of which is operated by Yorkshire Rider using new Scania single deck buses and the other in Ipswich marketed as SuperRoute 66 which is operated by Eastern Counties. Shown here is the official opening of the first part of the Leeds scheme. *Andrew Jarosz*

In 1994, the Group invested in major facilities in west Cornwall for Western National when a new depot was opened at Camborne. This was built on the site of the previous depot and has enabled centralisation of the maintenance facilities in the area. The cost of £670,000 was assisted by a Government Grant of £210,000.

FirstBus is active in research into new technology and intends to remain at the forefront of innervation in Public Transport. It has a specialised division, FirstBus Transit Developments, which progresses research and development of alternative fuels, guided busways and other transport related projects. Currently on trial in Bath are two vehicles powered by liquified petroleum gas (LPG) and work is also proceeding on the conversion of a vehicle using compressed natural gas (CNG) in Bristol.

In addition to the bus and coach operations, FirstBus wholly owns a number of other businesses including Badger Retail Limited (retail outlets in Bristol and Weston-super-Mare), FB Properties and has a majority interest in Skillplace Training Limited, based in South Wales and Black and White Design Limited of Leeds.

Firstbus has a policy of substantial investment in new vehicles and expects to take delivery of over 500 buses in the year to March 1996.

BADGERLINE

Badgerline Ltd, Enterprise House, Easton Road, Lawrence Hill, Bristol BS5 0DZ

Depots: Kensington Garage, London Road, Bath; Marlborough Street Bus Station, Bristol and Winterstoke Road Trading Estate, Weston-super-Mare. **Outstations:** Chepstow, Chippenham, Clevedon, Devizes, Dursley, Frome, Highbridge, Radstock, Trowbridge, Wells, Wotton-under-Edge and Yate

100-113

Volvo B10M-61   Alexander P   DP53F   1987

100	D100GHY	103	D103GHY	106	D106GHY	109	D109GHY	112	D112GHY
101	D101GHY	104	D104GHY	107	D107GHY	110	D110GHY	113	D113GHY
102	D102GHY	105	D105GHY	108	D108GHY	111	D111GHY		

121-136

Dennis Lance 11SDA3112   Plaxton Verde   B49F   1993-94

121	L121TFB	125	L125TFB	128	L128TFB	131	L131TFB	134	L134TFB
122	L122TFB	126	L126TFB	129	L129TFB	132	L132TFB	135	L135TFB
123	L123TFB	127	L127TFB	130	L130TFB	133	L133TFB	136	L136TFB
124	L124TFB								

137-142

Dennis Lance SLF 11SDA3201   Wright Pathfinder 320   B37F   1995   Seating variable

137	M137FAE	139	M139FAE	140	M140FAE	141	M141FAE	142	M142FAE
138	M138FAE							

201-225

Dennis Dart 9.8SDL3035   Plaxton Pointer   B40F   1993-94

201	L201SHW	206	L206SHW	211	L211VHU	216	L216VHU	221	L221VHU
202	L202SHW	207	L207SHW	212	L212VHU	217	L217VHU	223	L223VHU
203	L203SHW	208	L208SHW	213	L213VHU	218	L218VHU	224	L224VHU
204	L204SHW	209	L209SHW	214	L214VHU	219	L219VHU	225	L225VHU
205	L205SHW	210	L210VHU	215	L215VHU	220	L220VHU		

226-242

Dennis Dart 9.8SDL3054   Plaxton Pointer   B40F   1995

226	N226KAE	230	N230KAE	234	N234KAE	237	N237KAE	240	N240KAE
227	N227KAE	231	N231KAE	235	N235KAE	238	N238KAE	241	N241KAE
228	N228KAE	232	N232KAE	236	N236KAE	239	N239KAE	242	N242KAE
229	N229KAE	233	N233KAE						

290-296

Dennis Javelin GX 12SDA2153   Plaxton Expressliner 2   C49FT   1995

290	M290FAE	292	M292FAE	294	M294FAE	295	M295FAE	296	M296FAE
291	M291FAE	293	M293FAE						

2500	D500GHY	Volvo B10M-61	Van Hool Alizée	C49F	1987
2501	D501GHY	Volvo B10M-61	Van Hool Alizée	C49F	1987
2502	D502GHY	Volvo B10M-61	Van Hool Alizée	C49F	1987
2503	D503GHY	Volvo B10M-61	Van Hool Alizée	C49F	1987
2600	D600GHY	Volvo B10M-61	Van Hool Alizée	C57F	1987
2601	D601GHY	Volvo B10M-61	Van Hool Alizée	C57F	1987

3501-3534

Leyland National 2 NL116L11/1R   B52F*   1980   Ex Bristol, 1986
*3517 is DP52F, 3518-20 are DP47F, 3532/3 are DP49F.

3501	AAE645V	3511	AAE655V	3518	AAE662V	3527	BOU2V	3532	BOU7V
3508	AAE652V	3512	AAE656V	3519	AAE663V	3528	BOU3V	3533	BOU8V
3509	AAE653V	3513	AAE657V	3520	AAE664V	3529	BOU4V	3534	DHW349W
3510	AAE654V	3517	AAE661V	3526	BOU1V	3530	BOU5V		

Opposite, top: **The Dennis Lance was first introduced into the Badgerline fleet in 1993 with a delivery of sixteen examples with Plaxton Verde bodies. A similar combination has been allocated to several of the former Badgerline Group fleets since then. Seen heading for Bristol is 130, L130TFB.**
Bottom: **The delights of Cheddar Gorge are advertised on 8609, A809THW, a Leyland Olympian with convertible open-top bodywork by Roe. Shown here it is featuring the Cheddar Showcaves with part of the Gorge in the background.** *David Cole/Andrew Jarosz*

The latest buses for the Badgerline fleet are six Dennis Lance SLF, the super low floor version, fitted with Wright Endurance bodywork. Photographed on the Whiteway estate is 139, M139FAE.
Richard Godfrey

| 3535 | BVP819V | Leyland National 2 NL116L11/1R | | | B49F | 1980 | Ex Midland Red West, 1989 |
| 3536 | BVP820V | Leyland National 2 NL116L11/1R | | | B49F | 1980 | Ex Midland Red West, 1989 |

3601-3605

Leyland National 2 NL116HLXCT/1R — DP47F — 1984 — Ex Midland Red West, 1989

3601	A201YWP	3602	A202YWP	3603	A203YWP	3604	A204YWP	3605	A205YWP

3610-3616

Leyland Lynx LX2R11C15Z4R — Leyland Lynx — B49F — 1990

3610	H610YTC	3612	H612YTC	3614	H614YTC	3615	H615YTC	3616	H616YTC
3611	H611YTC	3613	H613YTC						

3800-3823

Mercedes-Benz 811D — Optare StarRider — B31F* — 1988 — *3810 is B33F
3819/21-3 are DP29F

3800	E800MOU	3804	E804MOU	3810	E810MOU	3815	E815MOU	3819	E819MOU
3801	E801MOU	3805	E805MOU	3811	E811MOU	3816	E816MOU	3821	E821MOU
3802	E802MOU	3806	E806MOU	3813	E813MOU	3817	E817MOU	3822	E822MOU
3803	E803MOU	3809	E809MOU	3814	E814MOU	3818	E818MOU	3823	E823MOU

| 3824 | F695AWW | Mercedes-Benz 811D | | Optare StarRider | B33F | 1988 | Ex Clapton Coaches, Radstock, 1994 |

3850-3866

Mercedes-Benz 709D — Reeve Burgess Beaver — B23F — 1991-92

3850	J850FTC	3854	J854FTC	3858	J858FTC	3861	J861HWS	3864	J864HWS
3851	J851FTC	3855	J855FTC	3859	J859FTC	3862	J862HWS	3865	J865HWS
3852	J852FTC	3856	J856FTC	3860	J860HWS	3863	J863HWS	3866	J866HWS
3853	J853FTC	3857	J857FTC						

3867-3876

Mercedes-Benz 709D — Plaxton Beaver — B23F — 1993

3867	K867NEU	3869	K869NEU	3871	K871NEU	3873	K873NEU	3875	K875NEU
3868	K868NEU	3870	K870NEU	3872	K872NEU	3874	K874NEU	3876	K876NEU

| 3877 | L877TFB | Mercedes-Benz 711D | | Plaxton Beaver | B23F | 1993 |

Minibus livery at Badgerline is now an all-over green with yellow lettering and a badger motif the only former Badgerline Group subsidiary to retain this feature. Twenty Mercedes-Benz 811Ds with Optare StarRider bodywork were delivered in 1988. In April 1994, as a result of the purchase of Clapton Coaches' stage services a similar vehicle arrived. Now numbered 3824, F695AWW, the vehicle was new in November 1988 to Dobson of Lostock Gralam when it carried high-back seating.
Phillip Stephenson

3878-3911

Mercedes-Benz 709D — Plaxton Beaver — B23F — 1993-94

3878	L878VHT	3885	L885VHT	3893	L893VHT	3899	L899VHT	3906 L906VHT
3879	L879VHT	3886	L886VHT	3894	L894VHT	3901	L901VHT	3907 L907VHT
3880	L880VHT	3887	L887VHT	3895	L895VHT	3902	L902VHT	3908 L908VHT
3881	L881VHT	3889	L889VHT	3896	L896VHT	3903	L903VHT	3909 L909VHT
3882	M882BEU	3890	L890VHT	3897	L897VHT	3904	L904VHT	3910 L910VHT
3883	L883VHT	3891	L891VHT	3898	L898VHT	3905	L905VHT	3911 L911VHT
3884	L884VHT	3892	L892VHT					

3912	E350AMR	Mercedes-Benz 609D	Reeve Burgess Beaver	DP25F	1988	Ex Clapton Coaches, Radstock, 1994
3913	F850TCW	Mercedes-Benz 609D	Reeve Burgess Beaver	B20F	1988	Ex Clapton Coaches, Radstock, 1994
3914	J850OBV	Mercedes-Benz 709D	Reeve Burgess Beaver	B23F	1992	Ex Clapton Coaches, Radstock, 1994
3915	L390UHU	Mercedes-Benz 709D	Plaxton Beaver	B23F	1993	Ex Clapton Coaches, Radstock, 1994
3916	K29OEU	Mercedes-Benz 709D	Wright Nim-Bus	B29F	1993	Ex Somerbus, Poulton, 1994
4448	B448WTC	Ford Transit 190	Dormobile	B16F	1985	On loan to Romania
4487	C487BFB	Ford Transit 190	Dormobile	B16F	1986	Ex Bristol, 1986
4703	C680ECV	Mercedes-Benz L608D	Reeve Burgess	B20F	1985	Ex Western National, 1993
4704	C681ECV	Mercedes-Benz L608D	Reeve Burgess	B20F	1985	Ex Western National, 1993

4705-4709

Mercedes-Benz L608D — PMT — B20F — 1986 — Ex Western National, 1994

4705	C979GCV	4706	C980GCV	4707	C981GCV	4708	C982GCV	4709 C983GCV

4710-4715

Mercedes-Benz L608D — Robin Hood — B20F — 1986 — Ex Western National, 1994

4710	C984GCV	4712	C986GCV	4713	C987GCV	4714	C988GCV	4715 C989GCV
4711	C985GCV							

4901-4947   Iveco Daily 49.10   Robin Hood City Nippy   B19F*   1986-87 *4907/43 are DP19F
4922-5/7-31/45/7 ex Midland Red West, 1990-94

4901	D901GEU	4909	D909HOU	4928	E928KEU	4934	E934KEU	4940	E940KEU
4902	D902HOU	4922	E922KEU	4929	E929KEU	4935	E935KEU	4942	E942LAE
4903	D903HOU	4923	E923KEU	4930	E930KEU	4936	E936KEU	4943	E943LAE
4904	D904HOU	4924	E924KEU	4931	E931KEU	4937	E937KEU	4945	E945LAE
4905	D905HOU	4925	E925KEU	4932	E932KEU	4939	E939KEU	4947	E952LAE
4906	D906HOU	4927	E927KEU						

4950	E209BOD	Iveco Daily 49.10	Reeve Burgess Beaver	B21F	1988	Ex Western National, 1992
4951	E210BOD	Iveco Daily 49.10	Reeve Burgess Beaver	B21F	1988	Ex Western National, 1992
4952	E211BOD	Iveco Daily 49.10	Reeve Burgess Beaver	B21F	1988	Ex Western National, 1992
4953	E212BOD	Iveco Daily 49.10	Reeve Burgess Beaver	B21F	1988	Ex Western National, 1992
4954	E208BOD	Iveco Daily 49.10	Reeve Burgess Beaver	B21F	1988	Ex Western National, 1992
5000	C28EUH	Leyland Olympian ONTL11/2R	East Lancashire	CH47/31F	1985	Ex G&G, Leamington, 1989
5001	C29EUH	Leyland Olympian ONTL11/2R	East Lancashire	CH47/31F	1985	Ex G&G, Leamington, 1989
5146	AHU523V	Bristol VRT/SL3/6LXB	Eastern Coach Works	DPH39/28F	1980	Ex Bristol, 1986
5501	HTC727N	Bristol VRT/SL2/6LXB	Eastern Coach Works	H39/31F	1975	Ex Bristol, 1986
5505	KOU791P	Bristol VRT/SL3/6LXB	Eastern Coach Works	H39/31F	1976	Ex Bristol, 1986
5506	KOU792P	Bristol VRT/SL3/6LXB	Eastern Coach Works	H39/31F	1976	Ex Bristol, 1986
5508	KOU794P	Bristol VRT/SL3/6LXB	Eastern Coach Works	H39/31F	1976	Ex Bristol, 1986
5509	KOU795P	Bristol VRT/SL3/6LXB	Eastern Coach Works	H39/31F	1976	Ex Bristol, 1986
5511	LHT721P	Bristol VRT/SL3/501(6LXB)	Eastern Coach Works	H39/31F	1976	Ex Bristol, 1986

5517-5526   Bristol VRT/SL3/6LXB   Eastern Coach Works   H43/31F*   1977-78 Ex Bristol, 1986
*5517/8/22 are DPH43/31F

5517	PEU512R	5519	PEU514R	5522	PEU517R	5523	PEU518R	5526	RFB616S
5518	PEU513R								

5529-5546   Bristol VRT/SL3/680*   Eastern Coach Works   H43/31F*   1981   Ex Bristol, 1986
*5529/33/4/7/41/4 now 6LXB; 5529/37/44 are DPH43/31F

5529	DHW351W	5534	EWS742W	5537	EWS745W	5542	EWS750W	5545	EWS753W
5531	EWS739W	5536	EWS744W	5541	EWS749W	5544	EWS752W	5546	EWS754W
5533	EWS741W								

Fifteen Volvo Citybus double-decks are used by Badgerline each being fitted with high-back seating. Twelve of the units were delivered new to the company in 1987, while three came from Western National. One of the latter, 5712, E215BTA, is seen in Bristol while operating express service X39 to Bath.
David Cole

5550	STW26W	Bristol VRT/SL3/6LXB	Eastern Coach Works	H39/31F	1981	Ex Thamesway, 1991
5551	KOO792V	Bristol VRT/SL3/6LXB	Eastern Coach Works	H39/31F	1980	Ex Thamesway, 1991
5553	KOO791V	Bristol VRT/SL3/6LXB	Eastern Coach Works	H39/31F	1980	Ex Thamesway, 1991
5554	KOO793V	Bristol VRT/SL3/6LXB	Eastern Coach Works	H39/31F	1980	Ex Thamesway, 1991
5555	LBD922V	Bristol VRT/SL3/6LXB	Eastern Coach Works	H43/31F	1980	Ex Thamesway, 1991
5556	FRP906T	Bristol VRT/SL3/6LX	Eastern Coach Works	H43/31F	1979	Ex Thamesway, 1991
5557	STW29W	Bristol VRT/SL3/6LXB	Eastern Coach Works	H39/31F	1981	Ex Western National, 1992
5558	STW31W	Bristol VRT/SL3/6LXC(6LXB)	Eastern Coach Works	H39/31F	1981	Ex Thamesway, 1992
5559	STW32W	Bristol VRT/SL3/6LXB	Eastern Coach Works	H39/31F	1981	Ex Thamesway, 1992
5560	STW33W	Bristol VRT/SL3/6LXB	Eastern Coach Works	H39/31F	1981	Ex Thamesway, 1992
5561	STW34W	Bristol VRT/SL3/6LXB	Eastern Coach Works	H39/31F	1981	Ex Thamesway, 1992
5562	XHK221X	Bristol VRT/SL3/6LXB	Eastern Coach Works	H43/31F	1981	Ex Thamesway, 1992
5563	XHK222X	Bristol VRT/SL3/6LXB	Eastern Coach Works	H43/31F	1981	Ex Thamesway, 1992
5564	XHK224X	Bristol VRT/SL3/6LXB	Eastern Coach Works	H43/31F	1981	Ex Thamesway, 1992
5565	XHK227X	Bristol VRT/SL3/6LXB	Eastern Coach Works	H43/31F	1981	Ex Thamesway, 1992
5600	JNU136N	Bristol VRT/SL2/6LX	Eastern Coach Works	H39/31F	1975	Ex Bristol, 1986
5614	MUA873P	Bristol VRT/SL3/6LX(LPG)	Eastern Coach Works	H43/31F	1975	Ex Western National, 1993
5615	MUA874P	Bristol VRT/SL3/6LX(LPG)	Eastern Coach Works	H43/31F	1975	Ex Bristol, 1986

5700-5711
Volvo Citybus B10M-50   Alexander RH   DPH47/35F   1987

5700	D700GHY	**5703**	D703GHY	**5706**	D706GHY	**5708**	D708GHY	**5710**	D710GHY
5701	D701GHY	**5704**	D704GHY	**5707**	D707GHY	**5709**	D709GHY	**5711**	D711GHY
5702	D702GHY	**5705**	D705GHY						

5712	E215BTA	Volvo Citybus B10M-50	Alexander RH	DPH47/35F	1988	Ex Western National, 1989
5713	E216BTA	Volvo Citybus B10M-50	Alexander RH	DPH47/35F	1988	Ex Western National, 1989
5714	E217BTA	Volvo Citybus B10M-50	Alexander RH	DPH47/35F	1988	Ex Western National, 1989
8583	GHT127	Bristol K5G	Eastern Coach Works	O33/26R	1941	Ex Bristol, 1990
8600	RTH931S	Bristol VRT/SL3/501(6LXB)	Eastern Coach Works	CO43/31F	1977	Ex SWT, 1991
8601	RTH932S	Bristol VRT/SL3/501	Eastern Coach Works	CO43/31F	1977	Ex SWT, 1991
8605	VDV143S	Bristol VRT/SL3/6LXB	Eastern Coach Works	CO43/31F	1978	Ex Western National, 1993
8606	VDV137S	Bristol VRT/SL3/6LXB	Eastern Coach Works	CO43/31F	1977	Ex Western National, 1990
8608	UFX860S	Bristol VRT/SL3/6LXB	Eastern Coach Works	CO43/31F	1977	Ex Bristol, 1986

8609-8614
Leyland Olympian ONLXB/1R   Roe   CO47/29F   1984   Ex Bristol, 1986

8609	A809THW	**8611**	A811THW	**8612**	A812THW	**8613**	A813THW	**8614**	A814THW
8610	A810THW							

8615	JHW107P	Bristol VRT/SL3/6LXB	Eastern Coach Works	O43/29F	1975	Ex Bristol, 1986
8616	JHW108P	Bristol VRT/SL3/6LXB	Eastern Coach Works	O43/29F	1975	Ex Bristol, 1986
8617	JHW109P	Bristol VRT/SL3/6LXB	Eastern Coach Works	O43/29F	1975	Ex Bristol, 1986
8619	JHW114P	Bristol VRT/SL3/6LXB	Eastern Coach Works	O43/29F	1976	Ex Bristol, 1986
8620	LEU256P	Bristol VRT/SL3/6LXB	Eastern Coach Works	O43/27D	1976	Ex Bristol, 1989
8621	LEU269P	Bristol VRT/SL3/6LXB	Eastern Coach Works	O43/27D	1976	Ex Bristol, 1990
8622	LEU263P	Bristol VRT/SL3/6LXB	Eastern Coach Works	O43/27D	1976	Ex City Line, 1993

9001-9010
Leyland Olympian ONCL10/1RZ Leyland   H47/31F*   1989   *9009/10 are DPH43/29F

9001	G901TWS	**9003**	G903TWS	**9005**	G905TWS	**9007**	G907TWS	**9009**	G909TWS
9002	G902TWS	**9004**	G904TWS	**9006**	G906TWS	**9008**	G908TWS	**9010**	G910TWS

9506-9532
Leyland Olympian ONLXB/1R   Roe   DPH43/29F   1982-83 Ex Bristol, 1986

9506	JHU905X	**9509**	JHU908X	**9511**	JHU910X	**9514**	JHU913X	**9531**	NTC130Y
9507	JHU906X	**9510**	JHU909X	**9512**	JHU911X	**9516**	LWS32Y	**9532**	NTC131Y
9508	JHU907X							

Previous Registrations:

530OHU	A205SAE	CSV303	A207SAE	CSV992	A209SAE
CSV231	A206SAE	CSV524	A208SAE		

Liveries: Yellow and emerald green; National Express 290-6, 2500-3; Guide Friday/The Bath Tour:8583, 8600/1/5/6/8/15-7/20/1.

Named Vehicles: 8583 *Prince Bladud*, 8600 *I.K.Brunel*, 8601 *William Herschel*, 8605 *Jane Austen*, 8606 *Minerva*, 8608 *John Wood*, 8615 *Sally Lunn*, 8616 *Beau Nash*, 8617 *Ralph Allen*, 8620 *King Edgar*, 8621 *Dr William Oliver*.

BREWERS

A E & F R Brewer Ltd, The Bus Garage, Acacia Avenue, Sandfields Estate,
Port Talbot SA12 7DW

Depots: Bridgend Industrial Estate, Bridgend; Alexandra Road, Gorseinon; Tumulus Way, Llandow Trading Estate, Llandow;
Heol Ty Gwyn Industrial Estate, Tyle Teg, Maesteg; Acacia Avenue, Sandfields Estate, Port Talbot;

117	C974GCV	Leyland Tiger TRCTL11/3RH	Duple 340	C48FT	1986	Ex Western National, 1995
118	MKH889A	Leyland Tiger TRCTL11/3R	Duple Caribbean	C46FT	1983	Ex United Welsh Coaches, 1992
119	C975GCV	Leyland Tiger TRCTL11/3RH	Duple 340	C48FT	1986	Ex Western National, 1995
120	MKH831A	Leyland Tiger TRCTL11/3R	Duple Caribbean	C46FT	1983	Ex United Welsh Coaches, 1992
121	MKH774A	Leyland Tiger TRCTL11/3R	Duple Caribbean	C46FT	1984	Ex United Welsh Coaches, 1992
122	C976GCV	Leyland Tiger TRCTL11/3RH	Duple 340	C48FT	1986	Ex Western National, 1995
123	C977GCV	Leyland Tiger TRCTL11/3RH	Duple 340	C48FT	1986	Ex Western National, 1995
124	MKH60A	Leyland Tiger TRCTL11/3R	Duple Caribbean	C48FT	1984	Ex United Welsh Coaches, 1990
126	MKH49A	Leyland Tiger TRCTL11/3R	Duple Caribbean	C46FT	1984	Ex SWT, 1992
127	MKH59A	Leyland Tiger TRCTL11/3RH	Duple Caribbean 2	C46FT	1984	Ex SWT, 1992
128	MKH69A	Leyland Tiger TRCTL11/3RH	Duple Caribbean 2	C46FT	1985	Ex SWT, 1993
129	MKH87A	Leyland Tiger TRCTL11/3RH	Duple Caribbean 2	C46FT	1985	Ex SWT, 1993
130	MKH98A	Leyland Tiger TRCTL11/3RH	Duple Caribbean 2	C46FT	1985	Ex SWT, 1994
138	300CUH	Hestair Duple SDA1512	Duple 425	C53FT	1988	Ex Western National, 1993
140	LIL5068	Hestair Duple SDA1512	Duple 425	C50FT	1988	Ex Western National, 1994
141	MKH48A	Leyland Royal Tiger B50	Roe Doyen	C50F	1983	Ex Midland Red West, 1992
142	RJI8032	Leyland Royal Tiger RT	Plaxton Paramount 3500	C49FT	1984	Ex United Welsh Coaches, 1992
144	JWE244W	Leyland Leopard PSU5D/4R	Plaxton Supreme IV	C49FT	1980	Ex Western National, 1995

145-151

		Leyland Leopard PSU5D/4R	Plaxton Supreme IV	C50F	1980-81	Ex United Welsh Coaches, 1992

145	JWE245W	**147**	JWE247W	**149**	JWE249W	**150**	JWE250W	**151**	JWE251W
146	JWE246W	**148**	JWE248W						

166	B221WEU	Leyland Tiger TRCTL11/3RH	Duple Laser 2	C47FT	1985	Ex Badgerline, 1994
167	VCL461	Leyland Tiger TRCTL11/3RH	Duple Laser 2	C47FT	1984	Ex Badgerline, 1994
170	K9BMS	Dennis Javelin 12SDA2117	Plaxton Premiére 320	C48FT	1993	
171	K13BMS	Dennis Javelin 12SDA2117	Plaxton Premiére 320	C48FT	1993	
172	L6BMS	Dennis Javelin 12SDA2125	Plaxton Premiére 320	C51FT	1994	
173	L8BMS	Dennis Javelin 12SDA2131	Plaxton Premiére 320	C51FT	1994	
174	L14BMS	Dennis Javelin 12SDA2131	Plaxton Premiére 350	C51FT	1994	
176	RJI8029	Leyland Tiger TRCTL11/3RZ	Plaxton Paramount 3500 II	C49FT	1985	Ex United Welsh Coaches, 1992
177	RJI8030	Leyland Tiger TRCTL11/3RZ	Plaxton Paramount 3500 II	C49FT	1985	Ex United Welsh Coaches, 1992
178	RJI8031	Leyland Tiger TRCTL11/3RZ	Plaxton Paramount 3500 II	C49FT	1985	Ex United Welsh Coaches, 1992
179	B905DHB	Leyland Tiger TRCTL11/3R	Duple Caribbean 2	C51FT	1985	Ex Cardiff Bus, 1995
180	B906DHB	Leyland Tiger TRCTL11/3R	Duple Caribbean 2	C51FT	1985	Ex Cardiff Bus, 1995

The Duple Laser had a short production life, only being built from 1982 to 1985. During that time two versions were built, an example of the later style with single-piece windscreen and single head-lamps is seen heading for Cardiff on service X1. Brewers 166, B221WEU, was new to Badgerline.
Byron Gage

Two vehicles from the Rhodeservice fleet have found their way into the Brewers fleet. Rhodeservice was acquired by Yorkshire Rider in 1994 but 506, G841PNW came straight into the Brewers fleet. This Van Hool A600 is unique in Britain though the type is to be found in large numbers, particularly in its home country, Belgium. *Cliff Beeton*

181	HHJ374Y	Leyland Tiger TRCTL11/2R	Alexander TE	C53F	1983	Ex Thamesway, 1991
182	HHJ377Y	Leyland Tiger TRCTL11/2R	Alexander TE	C49F	1983	Ex SWT, 1993
183	HHJ379Y	Leyland Tiger TRCTL11/2R	Alexander TE	C53F	1983	Ex Thamesway, 1991
184	A691OHJ	Leyland Tiger TRCTL11/2R	Alexander TE	C53F	1983	Ex Thamesway, 1991
185	A693OHJ	Leyland Tiger TRCTL11/2R	Alexander TE	C49F	1983	Ex SWT, 1993
186	A695OHJ	Leyland Tiger TRCTL11/2R	Alexander TE	C49F	1983	Ex SWT, 1993
188	278TNY	DAF MB230LT615	Plaxton Paramount 3500 III	C53FT	1989	Ex SWT, 1992
189	LIL5069	DAF MB200DKFL600	Plaxton Paramount 3500 II	C51FT	1986	Ex Western National, 1995
190	LIL5070	DAF MB230DKFL615	Duple 340	C48FT	1987	Ex Western National, 1995
191	LIL5071	DAF MB230DKFL615	Duple 340	C53FT	1987	Ex Western National, 1995
192	948RJO	DAF MB230LT615	Plaxton Paramount 3500 III	C51FT	1989	Ex United Welsh, 1992
196	972SYD	Scania K112CRB	Jonckheere Jubilee P50	C51FT	1988	Ex Cardiff Bus, 1995
197	F907DHB	Scania K113CRB	Plaxton Paramount 3500 III	C49FT	1989	Ex Cardiff Bus, 1995
198	YBK132	Leyland Royal Tiger RTC	Leyland Doyen	C49FT	1986	Ex United Welsh Coaches, 1992
199	605BBO	Leyland Royal Tiger RTC	Leyland Doyen	C49FT	1986	Ex United Welsh Coaches, 1992

201-214   Mercedes-Benz L608D   Robin Hood   B20F   1985-86 Ex SWT 1992-93

201	C201HTH	204	C204HTH	207	C207HTH	210	C210HTH	213	C213HTH
202	C202HTH	205	C205HTH	208	C208HTH	211	C211HTH	214	C214HTH
203	C203HTH	206	C206HTH	209	C209HTH	212	C212HTH		

216	D216LCY	Mercedes-Benz L608D	Robin Hood	DP19F	1986	Ex SWT 1993
217	D217LCY	Mercedes-Benz L608D	Robin Hood	DP19F	1986	Ex SWT 1993
218	WCY701	Mercedes-Benz L608D	Robin Hood	DP19F	1986	Ex SWT 1993
219	D219LCY	Mercedes-Benz L608D	Robin Hood	DP19F	1986	Ex SWT 1993
220	C209HJN	Mercedes-Benz L608D	Reeve Burgess	B20F	1985	Ex Eastern National, 1995
221	C221HJN	Mercedes-Benz L608D	Reeve Burgess	B20F	1985	Ex Eastern National, 1995
222	C478BHY	Mercedes-Benz L608D	Reeve Burgess	B20F	1986	Ex Eastern National, 1995
223	C480BHY	Mercedes-Benz L608D	Reeve Burgess	B20F	1986	Ex Eastern National, 1995
224	C481BHY	Mercedes-Benz L608D	Reeve Burgess	DP20F	1986	Ex Eastern National, 1995

225-229

Mercedes-Benz L608D | Dormobile | B20F | 1986 | Ex City Line, 1995

| 225 | D514FAE | 226 | D517FAE | 227 | D520FAE | 228 | D526FAE | 229 | D527FAE |

401-410

Mercedes-Benz 811D | Plaxton Beaver | B31F | 1993

| 401 | K401BAX | 403 | K403BAX | 405 | K405BAX | 407 | K407BAX | 409 | K409BAX |
| 402 | K402BAX | 404 | K404BAX | 406 | K406BAX | 408 | K408BAX | 410 | K410BAX |

411-418

Mercedes-Benz 709D | Reeve Burgess Beaver | B25F* | 1988 | Ex SWT, 1994; 411/3/7 are B23F

| 411 | F601AWN | 413 | F603AWN | 415 | F605AWN | 417 | F318AWN | 418 | F608AWN |
| 412 | F602AWN | 414 | F604AWN | 416 | F606AWN |

423-428

Mercedes-Benz 709D | Reeve Burgess Beaver | B25F* | 1988 | Ex SWT, 1994; *423/8 are B23F

| 423 | E283UCY | 424 | E284UCY | 425 | E285UCY | 426 | E286UCY | 428 | E288VEP |

506	G841PNW	Van Hool A600	Van Hool	B52F	1990	Ex Rhodeservice, Yeadon, 1994
507	G261LUG	Leyland Lynx LX112L10ZR1R	Leyland Lynx	B51F	1989	Ex Rhodeservice, Yeadon, 1994
508	J916WVC	Leyland Lynx LX2R11V18245	Leyland Lynx	B47F	1992	Ex Volvo demonstrator, 1992
509	J375WWK	Leyland Lynx LX2R11V18245	Leyland Lynx	B47F	1992	Ex Volvo demonstrator, 1992
510	K10BMS	Leyland Lynx LX2R11C15Z4A	Leyland Lynx	B47F	1992	
511	K11BMS	Leyland Lynx LX2R11C15Z4A	Leyland Lynx	B47F	1992	
512	K12BMS	Leyland Lynx LX2R11C15Z4A	Leyland Lynx	B47F	1992	

601-608

Dennis Dart 9.8SDL3035 | Plaxton Pointer | B40F | 1994

| 601 | L601FKG | 603 | L603FKG | 605 | L605FKG | 607 | L607FKG | 608 | L608FKG |
| 602 | L602FKG | 604 | L604FKG | 606 | L606FKG |

609-618

Dennis Dart 9.8SDL3054 | Plaxton Pointer | DP40F | 1995

| 609 | N609MHB | 611 | N611MHB | 613 | N613MHB | 615 | N615MHB | 617 | N617MHB |
| 610 | N610MHB | 612 | N612MHB | 614 | N614MHB | 616 | N616MHB | 618 | N618MHB |

| 716 | ERP558T | Leyland National 11351A/1R | | B49F | 1979 | Ex Badgerline, 1991 |

794-812

Leyland National 11351A/1R | | B49F* | 1976-79 | Ex SWT, 1989-1994
*803 is DP52F

| 794 | OEP794R | 801 | TWN801S | 806 | WWN806T | 810 | AWN810V | 812 | AWN812V |
| 795 | OEP795R | 803 | TWN803S | 809 | WWN809T |

946	XHK234X	Bristol VRT/SL3/6LXB	Eastern Coach Works	H43/31F	1981	Ex United Welsh Coaches, 1992
947	UAR587W	Bristol VRT/SL3/6LXB	Eastern Coach Works	H43/31F	1981	Ex United Welsh Coaches, 1992
948	UAR588W	Bristol VRT/SL3/6LXB	Eastern Coach Works	H43/31F	1981	Ex United Welsh Coaches, 1992
949	UAR598W	Bristol VRT/SL3/6LXB	Eastern Coach Works	H43/31F	1981	Ex United Welsh Coaches, 1992
959	WTH959T	Bristol VRT/SL3/501	Eastern Coach Works	H43/31F	1979	Ex SWT, 1992

971-988

Bristol VRT/SL3/501 | Eastern Coach Works | H43/31F | 1979-80 Ex SWT, 1992-94

| 971 | BEP971V | 974 | BEP974V | 981 | BEP981V | 985 | BEP985V | 988 | ECY988V |
| 972 | BEP972V | 977 | BEP977V | 982 | BEP982V | 986 | BEP986V |

Previous Registrations:

278TNY	F200EEP	LIL5071	E978WTA	MKH889A	RCY118Y
300CUH	E218CFJ	MKH48A	SOH555Y	MKH98A	B130CTH
605BBO	D401GHT	MKH49A	A259VWO, A126WEP	RJI8029	B34UNW, SWN159, 948RJO
948RJO	G500JEP	MKH59A	B127CTH	RJI8030	B35UNW, 300CUH
972SYD	E701NNH	MKH60A	A124XEP	RJI8031	B36UNW, WCY701
G841PNW	G680TKE, A6RLR	MKH69A	B128CTH	RJI8032	A840SYR, FDZ985
LIL5068	E207BOD	MKH774A	A255VWO, 605BBO	VCL461	B223WEU
LIL5069	C792MVH	MKH831A	RCY120Y	WCY701	D218LCY
LIL5070	E340WTT	MKH87A	B129CTH	YBK132	D400GHT

Liveries: Red, white and yellow (buses); White, red and green (Coaches); White (National Express) 122/7.

Opposite: **Deliveries to the Brewers fleet have seen sixteen Dennis Darts that continue to replace inherited Leyland Nationals. Photographed departing from Port Talbot bus station for Maesteg, is Dart 604, L604FKG. One of the final series 1 Nationals 812, AWN812V, was new to South Wales. It is seen working service 15, arriving in Bridgend.** *Cliff Beeton/John Jones*

CITY LINE

Bristol Omnibus Co Ltd, Enterprise House, Easton Road, Lawrence Hill,
Bristol BS5 0DZ

Depots: Winterstoke Road, Ashton, Bristol; Muller Road, Horfield, Bristol and Easton Road, Lawrence Hill, Bristol.

1501-1508
Dennis Dart 9SDL3034 — Plaxton Pointer — B35F — 1994

1501	L501VHU	**1503**	L503VHU	**1505**	L505VHU	**1507**	L507VHU	**1508**	L508VHU
1502	L502VHU	**1504**	L504VHU	**1506**	L506VHU				

1509-1547
Dennis Dart 9.8SDL3053 — Plaxton Pointer — DP35F — 1995

1509	M509DHU	**1517**	M517DHU	**1525**	M525FFB	**1533**	M533FFB	**1541**	N541HAE
1510	M510DHU	**1518**	M518DHU	**1526**	M526FFB	**1534**	M534FFB	**1542**	N542HAE
1511	M511DHU	**1519**	M519DHU	**1527**	M527FFB	**1535**	M535FFB	**1543**	N543HAE
1512	M512DHU	**1520**	M520FFB	**1528**	M528FFB	**1536**	M536FFB	**1544**	N544HAE
1513	M513DHU	**1521**	M521FFB	**1529**	M529FFB	**1537**	M537FFB	**1545**	N545HAE
1514	M514DHU	**1522**	M522FFB	**1530**	M530FFB	**1538**	M538FFB	**1546**	N546HAE
1515	M515DHU	**1523**	M523FFB	**1531**	M531FFB	**1539**	N539HAE	**1547**	N547HAE
1516	M516DHU	**1524**	M524FFB	**1532**	M532FFB	**1540**	N540HAE		

1548	N548HAE	Dennis Dart CNG unit	Plaxton Pointer	DP35F	1995	

1600-1662
Leyland Lynx LX2R11C15Z4R — Leyland Lynx — B49F — 1989-90

1600	F600RTC	**1613**	F613RTC	**1626**	F626RTC	**1639**	H639YHT	**1651**	H651YHT
1601	F601RTC	**1614**	F614RTC	**1627**	F627RTC	**1640**	H640YHT	**1652**	H652YHT
1602	F602RTC	**1615**	F615RTC	**1628**	F628RTC	**1641**	H641YHT	**1653**	H653YHT
1603	F603RTC	**1616**	F616RTC	**1629**	F629RTC	**1642**	H642YHT	**1654**	H654YHT
1604	F604RTC	**1617**	F617RTC	**1630**	F630RTC	**1643**	H643YHT	**1655**	H655YHT
1605	F605RTC	**1618**	F618RTC	**1631**	F631RTC	**1644**	H644YHT	**1656**	H656YHT
1606	F606RTC	**1619**	F619RTC	**1632**	F632RTC	**1645**	H645YHT	**1657**	H657YHT
1607	F607RTC	**1620**	F620RTC	**1633**	H633YHT	**1646**	H646YHT	**1658**	H658YHT
1608	F608RTC	**1621**	F621RTC	**1634**	H634YHT	**1647**	H647YHT	**1659**	H659YHT
1609	F609RTC	**1622**	F622RTC	**1636**	H636YHT	**1648**	H648YHT	**1660**	H660YHT
1610	F610RTC	**1623**	F623RTC	**1637**	H637YHT	**1649**	H649YHT	**1661**	H661YHT
1611	F611RTC	**1624**	F624RTC	**1638**	H638YHT	**1650**	H650YHT	**1662**	H662YHT
1612	F612RTC	**1625**	F625RTC						

5073-5151
Bristol VRT/SL3/6LXB — Eastern Coach Works — H43/27D — 1976-80

5073	MOU747R	**5140**	AHU517V	**5147**	AHW198V	**5149**	AHW200V	**5151**	AHW202V

7473	C473BHY	Ford Transit 190D	Dormobile	B16F	1986	

7570-7617
Iveco Daily 49.10 — Dormobile Routemaker — B20F — 1988-89

7570	E570NFB	**7580**	E580OOU	**7590**	F590OHT	**7600**	F600PWS	**7609**	F609PWS
7571	E571NFB	**7581**	E581OOU	**7591**	F591OHT	**7601**	F601PWS	**7610**	F610PWS
7572	E572NFB	**7582**	E582OOU	**7592**	F592OHT	**7602**	F602PWS	**7611**	F611PWS
7573	E573NFB	**7583**	E583OOU	**7593**	F593OHT	**7603**	F603PWS	**7612**	F612PWS
7574	E574NFB	**7584**	E584OOU	**7594**	F594OHT	**7604**	F604PWS	**7613**	F613PWS
7575	E575NFB	**7585**	E585OOU	**7595**	F595OHT	**7605**	F605PWS	**7614**	F614PWS
7576	E576NFB	**7586**	E586OOU	**7596**	F596OHT	**7606**	F606PWS	**7615**	F615PWS
7577	E577NFB	**7587**	E587OOU	**7597**	F597OHT	**7607**	F607PWS	**7616**	F616PWS
7578	F578OOU	**7588**	E588OOU	**7598**	F598PWS	**7608**	F608PWS	**7617**	F617PWS
7579	F579OOU	**7589**	E589OOU	**7599**	F599PWS				

Opposite, top: **In 1984 the Bristol Omnibus Company was split with Badgerline forming the country operation. The city operation was then sold, in 1987 to a joint management/Midland Red West team and the operation was later relaunched as City Line. Two types of single deck dominate, the Dart and the Leyland Lynx, of which some 63 are operated. Shown here is 1610, F610RTC.** *David Cole*
Opposite, bottom: **Six of the attractive Northern Counties Palatine II-bodied Volvo Olympians are in a silver, blue, and red livery for a park and ride service. Heading back to the Bath Road car-park, near to the old Bristol manufacturing plant is 9651, L651SEU.** *David Cole*

Waiting to enter a box junction are recently withdrawn 7500, the initial Mercedes-Benz L608D vehicle currently in the SWT fleet, and 9542, NTC141Y. This Leyland Olympian is one of many in the City Line fleet that carry high-bridge Roe bodywork. The LH code on the front panel indicates the vehicle is based at Lawrence Hill. *David Cole*

7801-7826

Mercedes-Benz 709D   Plaxton Beaver   B22F   1993

7801	L801SAE	7807	L807SAE	7812	L812SAE	7817	L817SAE	7822	L822SAE
7802	L802SAE	7808	L808SAE	7813	L813SAE	7818	L818SAE	7823	L823SAE
7803	L803SAE	7809	L809SAE	7814	L814SAE	7819	L819SAE	7824	L824SAE
7804	L804SAE	7810	L810SAE	7815	L815SAE	7820	L820SAE	7825	L825SAE
7805	L805SAE	7811	L811SAE	7816	L816SAE	7821	L821SAE	7826	L826SAE
7806	L806SAE								

7827-7874

Mercedes-Benz 709D   Plaxton Beaver   B22F   1994

7827	L827WHY	7837	M837ATC	7847	M847ATC	7857	M857ATC	7866	M866ATC
7828	L828WHY	7838	M838ATC	7848	M848ATC	7858	M858ATC	7867	M867ATC
7829	L829WHY	7839	M839ATC	7849	M849ATC	7859	M859ATC	7868	M868ATC
7830	L830WHY	7840	M840ATC	7850	M850ATC	7860	M860ATC	7869	M869ATC
7831	M831ATC	7841	M841ATC	7851	M851ATC	7861	M861ATC	7870	M870ATC
7832	M832ATC	7842	M842ATC	7852	M852ATC	7862	M862ATC	7871	M871ATC
7833	M833ATC	7843	M843ATC	7853	M853ATC	7863	M863ATC	7872	M872ATC
7834	M834ATC	7844	M844ATC	7854	M854ATC	7864	M864ATC	7873	M873ATC
7835	M835ATC	7845	M845ATC	7855	M855ATC	7865	M865ATC	7874	M874ATC
7836	M836ATC	7846	M846ATC	7856	M856ATC				

7875-7907

Mercedes-Benz 709D   Plaxton Beaver   B22F   1995

7875	N875HWS	7882	N882HWS	7889	N889HWS	7895	N895HWS	7902	N902HWS
7876	N876HWS	7883	N883HWS	7890	N890HWS	7896	N896HWS	7903	N903HWS
7877	N877HWS	7884	N884HWS	7891	N891HWS	7897	N897HWS	7904	N904HWS
7878	N878HWS	7885	N885HWS	7892	N892HWS	7898	N898HWS	7905	N905HWS
7879	N879HWS	7886	N886HWS	7893	N893HWS	7899	N899HWS	7906	N906HWS
7880	N880HWS	7887	N887HWS	7894	N894HWS	7901	N901HWS	7907	N907HWS
7881	N881HWS								

8138	D138NUS	Mercedes-Benz L608D	Alexander AM	B21F	1986	Ex Kelvin Central, 1992
8707	L707LKY	Mercedes-Benz 711D	Plaxton Beaver	B23F	1994	Ex Mercedes-Benz demonstrator, 1995
8880	M880ATC	Volvo B6-9M	Plaxton Pointer	B35F	1994	
8976	E976MFB	Iveco Daily 49.10	Dormobile Routemaker	B20F	1988	Ex Ford demonstrator, 1989

9501-9568 Leyland Olympian ONLXB/1R Roe H47/29F 1982-84

9501	JHU900X	**9530**	NTC129Y	**9541**	NTC141Y	**9551**	A951SAE	**9560**	A960THW
9502	JHU901X	**9534**	NTC133Y	**9543**	NTC142Y	**9552**	A952SAE	**9561**	A961THW
9503	JHU902X	**9535**	NTC134Y	**9544**	NTC143Y	**9553**	A953SAE	**9562**	A962THW
9504	JHU903X	**9536**	NTC135Y	**9545**	A945SAE	**9554**	A954SAE	**9563**	A963THW
9505	JHU904X	**9537**	NTC136Y	**9546**	A946SAE	**9555**	A955THW	**9564**	A964THW
9515	JHU914X	**9538**	NTC137Y	**9547**	A947SAE	**9556**	A956THW	**9565**	A965THW
9526	LWS42Y	**9539**	NTC138Y	**9548**	A948SAE	**9557**	A957THW	**9566**	A966THW
9527	LWS43Y	**9540**	NTC139Y	**9549**	A949SAE	**9558**	A958THW	**9567**	A967THW
9528	LWS44Y	**9541**	NTC140Y	**9550**	A950SAE	**9559**	A959THW	**9568**	A968THW
9529	LWS45Y								

9601-9630 Leyland Olympian ON2R56C16Z4 Northern Counties Palatine H44/32F 1992-93

9601	K601LAE	**9607**	K607LAE	**9613**	K613LAE	**9619**	K619LAE	**9625**	K625LAE
9602	K602LAE	**9608**	K608LAE	**9614**	K614LAE	**9620**	K620LAE	**9626**	K626LAE
9603	K603LAE	**9609**	K609LAE	**9615**	K615LAE	**9621**	K621LAE	**9627**	K627LAE
9604	K604LAE	**9610**	K610LAE	**9616**	K616LAE	**9622**	K622LAE	**9628**	K628LAE
9605	K605LAE	**9611**	K611LAE	**9617**	K617LAE	**9623**	K623LAE	**9629**	K629LAE
9606	K606LAE	**9612**	K612LAE	**9618**	K618LAE	**9624**	K624LAE	**9630**	K630LAE

9631-9654 Volvo Olympian YN2RV18Z4 Northern Counties Palatine II H47/29F 1993-94

9631	L631SEU	**9636**	L636SEU	**9641**	L641SEU	**9646**	L646SEU	**9651**	L651SEU
9632	L632SEU	**9637**	L637SEU	**9642**	L642SEU	**9647**	L647SEU	**9652**	L652SEU
9633	L633SEU	**9638**	L638SEU	**9643**	L643SEU	**9648**	L648SEU	**9653**	L653SEU
9634	L634SEU	**9639**	L639SEU	**9644**	L644SEU	**9649**	L649SEU	**9654**	L654SEU
9635	L635SEU	**9640**	L640SEU	**9645**	L645SEU	**9650**	L650SEU		

Liveries: Red, blue and yellow, Silver, blue and red (Park & Ride): 9649-54

Most of the midi-buses in the City Line fleet are Dennis Darts, but during 1994 a 9-metre B6 was added to the fleet for evaluation. Numbered 8880, M880ATC, the vehicle is seen with WE (Winterstoke Road) garage code. The B6 profile differs from the Dart having a higher window line to the rear of the vehicle.
Andrew Jarosz

Durbin Coaches operate in the Bristol area with a variety of high-specification coaches including Van Hool-bodied DAFs and two LAG Panoramics. Here we see a variety of the stock which are painted in a white and blue livery.

DURBIN COACHES

Durbin Coaches Ltd, Enterprise House, Easton Road, Bristol BS5 0DZ

Depots : Easton Road, Bristol; Lawrence Hill, Bristol and Station Road, Patchway

	HTC726N	Bristol VRT/SL2/6LX	Eastern Coach Works	H39/31F	1975	Ex Badgerline, 1992
	LEU262P	Bristol VRT/SL3/6LX	Eastern Coach Works	H43/27D	1976	Ex Teagle, Bristol, 1991
	URF668S	Bristol VRT/SL3/501(6LXB)	Eastern Coach Works	H43/31F	1978	Ex PMT, 1992
	WKO131S	Bristol VRT/SL3/6LXB	Eastern Coach Works	H43/31F	1978	Ex Maidstone & District, 1992
	RHT511S	Bristol VRT/SL3/6LXB	Eastern Coach Works	H43/27D	1978	Ex City Line, 1994
w	HVU521V	Ford R1114	Duple Dominant II	C53F	1980	Ex Axe Vale, Biddisham, 1989
	AHW201V	Bristol VRT/SL3/6LXB	Eastern Coach Works	H43/27D	1980	Ex City Line, 1994
	DFB672W	Leyland Leopard PSU5D/4R	Duple Dominant III	C57F	1981	Ex Peter Carol, Bristol, 1986
	JBW211Y	Leyland Leopard PSU5/2L	Plaxton Paramount 3200	C53F	1982	Ex Cheney, Banbury, 1994
	UWA580Y	Ford R1114	Duple Dominant IV Express	C53F	1983	Ex Avon CC, 1989
	LSU788	DAF MB200DKTL600	Plaxton Paramount 3200	C53F	1983	Ex Eastville, Bristol, 1993
	PJI5625	Aüwaerter Neoplan N122/3	Aüwaerter Skyliner	CH48/12CT	1983	Ex Clevedon Motorways, 1992
	RJI2723	Leyland Tiger TRCTL11/3R	Plaxton Paramount 3500	C49F	1983	Ex Canavan, Kilsyth, 1991
	A105EBC	Ford R1115	Plaxton Paramount 3200	C53F	1983	Ex Glenvic, Bristol, 1991
	A611XKU	Leyland Tiger TRCTL11/3R	Plaxton Paramount 3200	C57F	1983	Ex Godson, Leeds, 1992
	WYY752	Bova EL28/581	Duple Calypso	C53F	1984	Ex Glenvic, Bristol, 1995
	A455JJF	DAF MB200DKFL600	Duple Caribbean	C55F	1984	Ex Eastville, Bristol, 1992
	SSU437	DAF MB200DKFL600	Plaxton Paramount 3500	C53F	1984	Ex Eastville, Bristol, 1990
	RJI2720	DAF SB2300DHS585	Plaxton Paramount 3500	C53F	1984	Ex Slack, Tansley, 1993
	RJI2721	Bova EL28/581	Duple Calypso	C53F	1984	Ex Cropper, Pudsey, 1993
	ESU980	Aüwaerter Neoplan N122/3	Aüwaerter Skyliner	CH57/20CT	1985	Ex Martindale, Ferryhill, 1994
	HXI578	Aüwaerter Neoplan N122/3	Aüwaerter Skyliner	CH53/20CT	1986	Ex Harris, Bromsgrove, 1995
	C305AHP	DAF SB2300DHS585	Duple 340	C57F	1986	Ex Travelfar, Henfield, 1991
	C468BHY	Ford Transit 190D	Dormobile	B16F	1986	Ex City Line, 1995
w	LBZ2303	Volvo B10M-61	Plaxton Paramount 3500 III	C49F	1986	Ex Clevedon Motorways, 1994
	D501FAE	Mercedes-Benz L608D	Dormobile	B20F	1986	Ex City Line, 1995
	D502FAE	Mercedes-Benz L608D	Dormobile	B20F	1986	Ex City Line, 1995
	D503FAE	Mercedes-Benz L608D	Dormobile	B20F	1986	Ex City Line, 1995
	D508FAE	Mercedes-Benz L608D	Dormobile	B20F	1986	Ex City Line, 1995
	D509FAE	Mercedes-Benz L608D	Dormobile	B20F	1986	Ex City Line, 1995
	D510FAE	Mercedes-Benz L608D	Dormobile	B20F	1986	Ex City Line, 1995
	D144NDT	Mercedes-Benz L608D	Whittaker	B20F	1986	Ex ?, 1994
	E694UND	Mercedes-Benz 609D	Made-to-Measure	B21F	1987	Ex Baker, Weston-super-Mare, 1994
	LBZ2305	DAF SB2305DHS585	Van Hool Alizée	C53F	1987	Ex London Coaches, 1994
	LBZ2571	DAF SB3000DKV601	Van Hool Alizée	C53F	1988	Ex Bennett, Gloucester, 1993
	LBZ2955	DAF SB3000DKV601	Van Hool Alizée	C51F	1988	Ex Chartercoach, Dovercourt, 1993
	RJI5704	LAG G355Z	LAG Panoramic	C49FT	1988	Ex Majestic, Shareshill, 1995
	RJI5706	LAG G355Z	LAG Panoramic	C49F	1989	Ex Majestic, Shareshill, 1995
	FBC1C	Van Hool T815	Van Hool Alizée	C49F	1989	Ex Penniston, Melton Mowbray, 1994

Previous Registrations:

ESU980	B380PAJ		LBZ2571	E341EVH	RJI2723	FWH39Y
FBC1C	F547TJF		LBZ2955	E354EVH	RJI5704	E135KRP
HXI578	C722JTL, FIL7287, C386DVG	LSU788	ANA436Y	RJI5706	F23WNH
JBW211Y	CJX563Y, URT682	PJI5625	MVL610Y	SSU437	A985UFB
LBZ2303	D818SGB		RJI2720	A463HJF	WYY752	A322HFP
LBZ2305	E312EVH		RJI2721	B799MAY		

Livery: Orange (buses and some minibuses); white and blue.

EASTERN COUNTIES

Eastern Counties Omnibus Co Ltd, 79 Thorpe Road, Norwich, NR1 1UA

Depots: Cotton Lane, Bury St Edmunds; Wellington Road, Great Yarmouth; Star Lane, Ipswich; Vancouver Avenue, King's Lynn; St Michael's Road, Kings Lynn; Gas Works Road, Lowestoft; Roundtree Way Norwich and Vulcan Road, Norwich.

OC1-5   Leyland Olympian ONLXB/1RZ   Northern Counties   H40/35F*   1989   *4 & 5 are DPH40/25F

1	F101AVG	2	F102AVG	3	F103AVG	4	F104AVG	5	F105AVG

JD6-10   Dennis Javelin 11SDL1933   Duple 300   DP48F   1989

6	G706JAH	7	G707JAH	8	G708JAH	9	G709JAH	10	G710JAH

JP11-20   Dennis Javelin 11SDL1924   Plaxton Derwent II   DP51F   1990

11	H611RAH	13	H613RAH	15	H615RAH	17	H617RAH	19	H619RAH
12	H612RAH	14	H614RAH	16	H616RAH	18	H618RAH	20	H620RAH

OL21-25   Leyland Olympian ON2R50G13Z4   Leyland   H45/31F   1991

21	J621BVG	22	J622BVG	23	J623BVG	24	J624BVG	25	J625BVG

MD26-35   Mercedes-Benz 609D   Dormobile   B20F   1992-93

26	K26HCL	28	K28HCL	30	J530FCL	32	K732JAH	34	K734JAH
27	K27HCL	29	K29HCL	31	K731JAH	33	K733JAH	35	K735JAH

LC36-40   Dennis Lance 11SDA3101   Northern Counties Paladin   B49F   1993

36	K736JAH	37	K737JAH	38	K738JAH	39	K739JAH	40	K740JAH

DP41	K741JAH	Dennis Dart 9SDL3011	Plaxton Pointer	B33F	1993
DP42	K742JAH	Dennis Dart 9SDL3011	Plaxton Pointer	B33F	1993
DP43	K743JAH	Dennis Dart 9SDL3011	Plaxton Pointer	B33F	1993
DP44	K744JAH	Dennis Dart 9SDL3011	Plaxton Pointer	B33F	1993

MG45-74   Mercedes-Benz 609D   Frank Guy   B20F   1993-94

45	L245PAH	51	L251PAH	57	L257PAH	63	M363XEX	70	M370XEX
46	L246PAH	52	L252PAH	58	L258PAH	64	M364XEX	71	M371XEX
47	L247PAH	53	L253PAH	59	L259PAH	65	M365XEX	72	M372XEX
48	L248PAH	54	L254PAH	60	M360XEX	67	M367XEX	73	M373XEX
49	L249PAH	55	L255PAH	61	M361XEX	68	M368XEX	74	M374XEX
50	L250PAH	56	L256PAH	62	M362XEX	69	M369XEX		

DP75-80   Dennis Dart 9SDL3041   Plaxton Pointer   B35F   1994

75	M375YEX	77	M377YEX	78	M378YEX	79	M379YEX	80	M380YEX
76	M376YEX								

IC81	F702MBC	Iveco Daily 49-10	Carlyle Dailybus 2	B25F	1988	Ex Leicester City Bus, 1994
IC82	F705MBC	Iveco Daily 49-10	Carlyle Dailybus 2	B25F	1988	Ex Leicester City Bus, 1994
IC83	F710NJF	Iveco Daily 49-10	Carlyle Dailybus 2	B25F	1988	Ex Leicester City Bus, 1994

VP84-93   Volvo B6-9.9M   Plaxton Pointer   B40F   1994

84	M584ANG	86	M586ANG	88	M588ANG	90	M590ANG	92	M592ANG
85	M585ANG	87	M587ANG	89	M589ANG	91	M591ANG	93	M593ANG

IC94-104   Iveco Daily 49-10   Carlyle Dailybus 2   B25F   1988-89 Ex Leicester City Bus, 1994-95

94	F708MBC	97	F701MBC	99	F704MBC	101	F707MBC	103	F716PFP
95	F706MBC	98	F715PFP	100	F709NJF	102	F719PFP	104	F722PFP
96	F703MBC								

Eastern Counties' new livery is almost a reverse of the scheme used prior to becoming part of GRT. The attractive scheme uses a cream base with red and orange house colours. To illustrate this are OC5, F105AVG, a Northern Counties-bodied Olympian with high-back seating and VP89, M589ANG, a Dennis Dart with Plaxton Pointer bodywork. *David Longbottom/Richard Godfrey*

Eastern Counties still have a considerable number of series 2 VRTs in stock but have gathered some later series 3s from elsewhere. Shown in Ipswich is VR186, ODL659R, acquired in 1991 from Southern Vectis. *Richard Godfrey*

MG105-124

| | | | | | | | | | Mercedes-Benz 609D | Frank Guy | B20F | 1995 |

105	N605GAH	**109**	N609GAH	**113**	N613GAH	**117**	N617GAH	**121**	N621GAH
106	N606GAH	**110**	N610GAH	**114**	N614GAH	**118**	N618GAH	**122**	N622GAH
107	N607GAH	**111**	N611GAH	**115**	N615GAH	**119**	N619GAH	**123**	N623GAH
108	N608GAH	**112**	N612GAH	**116**	N616GAH	**120**	N620GAH	**124**	N624GAH

DP125	N625GAH	Dennis Dart 9SDL3041	Plaxton Pointer	B35F	1995
DP126	N626GAH	Dennis Dart 9SDL3041	Plaxton Pointer	B35F	1995
MA127	N627GAH	Mercedes-Benz 709D	Plaxton Beaver	B25F	1995
MA128	N628GAH	Mercedes-Benz 709D	Plaxton Beaver	B25F	1995

VR129-160

| | | | | | | | | | Bristol VRT/SL2/6LX | Eastern Coach Works | H43/31F | 1974-75 |

129	RAH129M	**139**	SNG439M	**147**	GNG713N	**153**	JNG51N	**158**	JNG56N
134	RAH134M	**141**	TAH554N	**148**	GNG714N	**154**	JNG52N	**159**	JNG57N
136	SNG436M	**142**	GNG708N	**149**	GNG715N	**156**	JNG54N	**160**	JNG58N
138	SNG438M	**143**	GNG709N	**151**	JNG49N				

VR161	JNU137N	Bristol VRT/SL2/6LX	Eastern Coach Works	H43/31F	1975	Ex Western National, 1992
VR162	MCL938P	Bristol VRT/SL3/6LXB	Eastern Coach Works	H43/31F	1976	
VR163	OUP683P	Bristol VRT/SL3/6LXB	Eastern Coach Works	H43/31F	1976	Ex Western National, 1992

VR164-171

| | | | | | | | | | Bristol VRT/SL3/6LXB | Eastern Coach Works | H43/31F | 1976 |

164	MCL940P	**168**	MEX770P	**169**	MCL944P	**170**	MEX769P	**171**	MEX768P
165	MCL941P								

VL172	NAH135P	Bristol VRT/SL3/6LX(501)	Eastern Coach Works	H43/31F	1976	
VR176	NAH139P	Bristol VRT/SL3/6LXB	Eastern Coach Works	H43/31F	1976	
VR178	NAH141P	Bristol VRT/SL3/6LXB	Eastern Coach Works	H43/31F	1976	
VR179	OEL233P	Bristol VRT/SL3/501	Eastern Coach Works	H43/31F	1976	Ex Wilts & Dorset, 1993
VR180	OEL236P	Bristol VRT/SL3/501	Eastern Coach Works	H43/31F	1976	Ex Wilts & Dorset, 1993
VR181	OPW181P	Bristol VRT/SL3/6LXB	Eastern Coach Works	H43/31F	1976	
VL183	WDM345R	Bristol VRT/SL3/501	Eastern Coach Works	H43/31F	1977	Ex PMT, 1992
VR184	ODL657R	Bristol VRT/SL3/6LXB	Eastern Coach Works	H43/31F	1977	Ex Southern Vectis, 1991

| VR185 | ODL658R | Bristol VRT/SL3/6LXB | Eastern Coach Works | H43/31F | 1977 | Ex Southern Vectis, 1991 |
| VR186 | ODL659R | Bristol VRT/SL3/6LXB | Eastern Coach Works | H43/31F | 1977 | Ex Southern Vectis, 1991 |

VR187-211
Bristol VRT/SL3/6LXB   Eastern Coach Works   H43/31F   1976-78

187	PVF359R	192	TEX402R	197	TEX407R	203	XNG203S	206	XNG206S
188	PVF360R	193	TEX403R	198	TEX408R	204	XNG204S	207	XNG207S
189	RPW189R	194	TEX404R	199	WPW199S	205	XNG205S	211	YNG211S
191	TEX401R	196	TEX406R						

| VL215 | BRF691T | Bristol VRT/SL3/501 | Eastern Coach Works | H43/31F | 1978 | Ex PMT, 1993 |

VR216-236
Bristol VRT/SL3/6LXB   Eastern Coach Works   H43/31F   1978-79

216	BCL216T	219	BVG219T	222	BVG222T	225	BVG225T	230	DEX230T
217	BCL217T	220	BVG220T	223	BVG223T	226	DEX226T	236	DNG236T
218	BVG218T	221	BVG221T	224	BVG224T	229	DEX229T		

| VR237 | GRA844V | Bristol VRT/SL3/6LXB | Eastern Coach Works | H43/31F | 1980 | Ex Trent, 1991 |

VR238-244
Bristol VRT/SL3/6LXB   Eastern Coach Works   H43/31F   1979

| 238 | HAH238V | 240 | HAH240V | 242 | JAH242V | 243 | JAH243V | 244 | JAH244V |
| 239 | HAH239V | 241 | JAH241V | | | | | | |

VR245-250
Bristol VRT/SL3/6LXB   Eastern Coach Works   H43/31F   1980   Ex Trent, 1991

| 245 | GRA841V | 247 | GRA843V | 248 | GRA845V | 249 | GRA847V | 250 | GRA846V |
| 246 | GRA842V | | | | | | | | |

VR251-282
Bristol VRT/SL3/6LXB   Eastern Coach Works   H43/31F   1980-81

251	PCL251W	256	PCL256W	262	RAH262W	270	RAH270W	275	TAH275W
252	PCL252W	257	PCL257W	263	RAH263W	271	TAH271W	276	TAH276W
253	PCL253W	258	RAH258W	266	RAH266W	272	TAH272W	277	VAH277X
254	PCL254W	259	RAH259W	267	RAH267W	273	TAH273W	281	VAH281X
255	PCL255W	261	RAH261W	269	RAH269W	274	TAH274W	282	VAH282X

VR283-302
Bristol VRT/SL3/6LXB   Eastern Coach Works   H43/31F*   1981-82   *284/5/7 are DPH41/25F

283	VEX283X	286	VEX286X	288	VEX288X	292	VEX292X	297	VEX297X
284	VEX284X	287	VEX287X	290	VEX290X	294	VEX294X	302	VEX302X
285	VEX285X								

VR303-310
Bristol VRT/SL3/6LXB   Eastern Coach Works   H43/31F*   1981   Ex Trent, 1991

| 303 | PRC848X | 305 | PRC851X | 307 | PRC853X | 309 | PRC855X | 310 | PRC857X |
| 304 | PRC850X | 306 | PRC852X | 308 | PRC854X | | | | | |

Passing through the ford in Kersley is LC36, K736JAH. The Dennis Lance is one of five delivered in 1993 and fitted with Northern Counties Paladin bodywork. It was seen while running service 156, a special Suffolk Bus Sunday service.
David Cole

HVR331	KKE731N	Bristol VRT/SL2/6LX	Eastern Coach Works	H43/34F	1975	Ex Hastings & District, 1985
HVR332	KKE732N	Bristol VRT/SL2/6LX	Eastern Coach Works	H43/34F	1975	Ex Hastings & District, 1985
HVR333	KKE733N	Bristol VRT/SL2/6LX	Eastern Coach Works	H43/34F	1975	Ex Hastings & District, 1985
HVR334	KKE734N	Bristol VRT/SL2/6LX	Eastern Coach Works	H43/34F	1975	Ex Hastings & District, 1985
OT351	OCK995K	Bristol VRT/SL2/6LX	Eastern Coach Works	O39/31F	1972	Ex Ribble, 1985
OT352	NCK980J	Bristol VRT/SL2/6LX	Eastern Coach Works	O39/31F	1971	Ex Ribble, 1985
OT353	JNG50N	Bristol VRT/SL2/6LX	Eastern Coach Works	O43/31F	1975	
VR378	OCK988K	Bristol VRT/SL2/6LX	Eastern Coach Works	H39/31F	1972	Ex Ribble, 1985
VR384	OCK994K	Bristol VRT/SL2/6LX	Eastern Coach Works	H39/31F	1972	Ex Ribble, 1985
VR385	OCK985K	Bristol VRT/SL2/6LX	Eastern Coach Works	H39/31F	1972	Ex Ribble, 1985
VR410	CJO470R	Bristol VRT/SL3/6LX	Eastern Coach Works	H43/31F	1977	Ex City of Oxford, 1984
VR411	CJO471R	Bristol VRT/SL3/6LX	Eastern Coach Works	H43/31F	1977	Ex City of Oxford, 1984
VR412	CJO472R	Bristol VRT/SL3/6LX	Eastern Coach Works	H43/31F	1977	Ex City of Oxford, 1984

RA421-426

		Renault-Dodge S56	Alexander AM	B21F	1986	Ex SMT, 1994

421	D415ASF	**423**	D430ASF	**424**	D401ASF	**425**	D425ASF	**426**	D407ASF
422	D417ASF								

VR501	GAG48N	Bristol VRT/SL2/6LX	Eastern Coach Works	H43/31F	1974	Ex Rosemary, Terrington St Clement, 1993
VR502	YHN654M	Bristol VRT/SL2/6LX	Eastern Coach Works	H39/31F	1974	Ex Rosemary, Terrington St Clement, 1993
DD504	THX573S	Leyland Fleetline FE30ALRSp	Park Royal	H44/27D	1978	Ex Rosemary, Terrington St Clement, 1993
DD505	OJD195R	Leyland Fleetline FE30AGR	MCW	H45/32F	1977	Ex Rosemary, Terrington St Clement, 1993
DD506	THX531S	Leyland Fleetline FE30ALRSp	Park Royal	H44/27D	1978	Ex Rosemary, Terrington St Clement, 1993
DD507	WWH26L	Daimler Fleetline CRG6LX	Park Royal	H43/32F	1973	Ex Rosemary, Terrington St Clement, 1993
DD514	OSG74V	Leyland Fleetline FE30AGR	Eastern Coach Works	H43/32F	1979	Ex SMT, 1995
DD515	OSG55V	Leyland Fleetline FE30AGR	Eastern Coach Works	H43/32F	1979	Ex SMT, 1995
DD516	GSC856T	Leyland Fleetline FE30AGR	Eastern Coach Works	H43/32F	1978	Ex SMT, 1995
DD517	GSC857T	Leyland Fleetline FE30AGR	Eastern Coach Works	H43/32F	1978	Ex SMT, 1995
LP523	6920MX	Leyland Leopard PSU3B/4R	Plaxton Elite III Express	C51F	1974	Ex Rosemary, Terrington St Clement, 1993
BD525	EHE234V	Bedford YMT	Duple Dominant II	C53F	1980	Ex Rosemary, Terrington St Clement, 1993
LD527	GRF267V	Leyland Leopard PSU3E/4R	Duple Dominant II	C53F	1979	Ex Sanders, Holt, 1993
TP528	7694VC	Leyland Tiger TRCTL11/3R	Plaxton Paramount 3200 E	C53F	1983	Ex Vanguard, 1993
TP529	6149KP	Leyland Tiger TRCTL11/3R	Plaxton Paramount 3200 E	C53F	1983	Ex Vanguard, 1993
LD530	YFV179R	Leyland Leopard PSU3E/4R	Duple Dominant	C51F	1977	Ex Powell, Lapford, 1993
TA531	A692OHJ	Leyland Tiger TRCTL11/3R	Alexander TE	C53F	1983	Ex Eastern National, 1995
TA532	HHJ372Y	Leyland Tiger TRCTL11/3R	Alexander TE	C53F	1983	Ex Eastern National, 1995
LG567	PVF367R	Leyland National 11351A/1R	East Lancs Greenway (1995)	B52F	1976	
LG568	PVF368R	Leyland National 11351A/1R	East Lancs Greenway (1994)	B52F	1976	
LG569	PVF369R	Leyland National 11351A/1R	East Lancs Greenway (1995)	B52F	1976	
LG585	TVF620R	Leyland National 11351A/1R	East Lancs Greenway (1994)	B52F	1977	
LN586	WAH586S	Leyland National 11351A/1R		B49F	1977	
LG587	WAH587S	Leyland National 11351A/1R	East Lancs Greenway (1995)	B49F	1977	
LG588	WAH588S	Leyland National 11351A/1R	East Lancs Greenway (1995)	B52F	1977	
LN589	WAH589S	Leyland National 11351A/1R		B52F	1977	
LG590	WAH590S	Leyland National 11351A/1R	East Lancs Greenway (1995)	B52F	1977	
LN591	WAH591S	Leyland National 11351A/1R		B52F	1977	
LN592	WAH592S	Leyland National 11351A/1R		B52F	1977	
LG593	WAH593S	Leyland National 11351A/1R	East Lancs Greenway (1993)	B52F	1977	
LN594	WAH594S	Leyland National 11351A/1R		B52F	1977	
LG598	WVF598S	Leyland National 11351A/1R	East Lancs Greenway (1993)	B52F	1978	
LG599	WVF599S	Leyland National 11351A/1R	East Lancs Greenway (1995)	B52F	1978	

LN601-617

		Leyland National 2 NL116L11/1R		B49F	1980

601	KVG601V	**604**	KVG604V	**608**	KVG608V	**613**	PEX613W	**616**	PEX616W
602	KVG602V	**606**	KVG606V	**609**	KVG609V	**614**	PEX614W	**617**	PEX617W
603	KVG603V	**607**	KVG607V	**610**	PEX610W	**615**	PEX615W		

LN624-628

		Leyland National 2 NL116AL11/1R		B49F	1981

624	UVF624X	**625**	UVF625X	**626**	UVF626X	**627**	UVF627X	**628**	UVF628X

LN629	KEP829X	Leyland National 2 NL116L11/1R		B49F	1980	Ex Eastern National, 1995
SD637	C637BEX	Freight Rover Sherpa	Dormobile	B16F	1986	
SD638	C638BEX	Freight Rover Sherpa	Dormobile	B16F	1986	
SD652	C652BEX	Freight Rover Sherpa	Dormobile	B16F	1986	
MH701	E701TNG	Mercedes-Benz 609D	Robin Hood	B20F	1988	
MH702	E702TNG	Mercedes-Benz 609D	Robin Hood	B20F	1988	

MA711-728

						Mercedes-Benz L608D	Alexander AM	B20F	1986		

711	C711BEX	715	C715BEX	719	C719BEX	722	C722BEX	725	C725BEX
712	C712BEX	716	C716BEX	720	C720BEX	723	C723BEX	726	C726BEX
713	C713BEX	717	C717BEX	721	C721BEX	724	C724BEX	727	C727BEX
714	C714BEX	718	C718BEX						

MB741-757

Mercedes-Benz L608D   Reeve Burgess   B20F   1986

741	C741BEX	745	C745BEX	749	C749BEX	752	C752BEX	755	C755BEX
742	C742BEX	746	C746BEX	750	C750BEX	753	C753BEX	756	C756BEX
743	C743BEX	747	C747BEX	751	C751BEX	754	C754BEX	757	C757BEX
744	C744BEX	748	C748BEX						

MB758	D758LEX	Mercedes-Benz 609D		B20F	1987	
MB759	D759LEX	Mercedes-Benz 609D	Reeve Burgess	B20F	1987	
LN762u	XNG762S	Leyland National 11351A/1R		B49F	1978	
LN763	XNG763S	Leyland National 11351A/1R		B52F	1978	
LG765	XNG765S	Leyland National 11351A/1R	East Lancs Greenway (1995)	B52F	1978	
LN766u	XNG766S	Leyland National 11351A/1R		B52F	1978	
LN767	XNG767S	Leyland National 11351A/1R		B52F	1978	
LG768	XNG768S	Leyland National 11351A/1R	East Lancs Greenway (1995)	B52F	1978	

LN769-778

Leyland National 11351A/1R   B52F   1978   778 is B49F

LN769	XNG769S	LN774	CCL774T	LN775	CCL775T	LN776	CCL776T	LN778	CCL778T
LN770	XNG770S								

LG781	DPW781T	Leyland National 11351A/1R (6HLX) East Lancs Greenway (1994) B49F	1978	
LG782	DPW782T	Leyland National 11351A/1R	East Lancs Greenway (1994) B49F	1978	
LG783	VFX981S	Leyland National 11351A/1R	East Lancs Greenway (1994) B52F	1978	Ex Stagecoach South, 1993
LG784	UFX854S	Leyland National 11351A/1R	East Lancs Greenway (1994) B52F	1978	Ex Stagecoach South, 1993
LG785	YFY7M	Leyland National 11351A/1R	East Lancs Greenway (1994) B52F	1978	Ex Merseybus, 1993
LD791	XWX181S	Leyland Leopard PSU3E/4R	Duple Dominant II	C51F	1978	Ex Powell, Lapford, 1994
LW792	OEX792W	Leyland Leopard PSU3E/4R	Willowbrook 003	C49F	1980	Ex Ambassador Travel, 1987
LW793	OEX793W	Leyland Leopard PSU3E/4R	Willowbrook 003	C49F	1980	Ex Ambassador Travel, 1987
LW794	OEX794W	Leyland Leopard PSU3E/4R	Willowbrook 003	C49F	1980	Ex Ambassador Travel, 1987
LD795	LLT345V	Leyland Leopard PSU3E/4R	Duple Dominant II	C51F	1980	Ex Lewis, Whitwick, 1993
LW796	OEX796W	Leyland Leopard PSU3E/4R	Willowbrook 003	C49F	1980	Ex Ambassador Travel, 1987
LD799	RRB116R	Leyland Leopard PSU3E/4R	Duple Dominant I	C49F	1977	Ex Trent, 1993
LW808	JCL808V	Leyland Leopard PSU3E/4R	Willowbrook 003	C49F	1980	Ex Ambassador Travel, 1987
LW809	JCL809V	Leyland Leopard PSU3E/4R	Willowbrook 003	C49F	1980	Ex Ambassador Travel, 1987
TD850	D779NUD	Ford Transit VE6	Dormobile	B16F	1986	Ex East Kent, 1988

TH851-856

Ford Transit VE6   Robin Hood   B16F   1987

851	E851PEX	853	E853PEX	854	E854PEX	855	E855PEX	856	E856PEX
852	E852PEX								

TH892-919

Ford Transit 190D   Robin Hood   B16F   1986

892	C892BEX	896	C896BEX	903	C903BEX	908	C908BEX	913	C913BEX
894	C894BEX	897	C897BEX	905	C905BEX	912	C912BEX	919	C919BEX
895	C895BEX	898	C898BEX	906	C906BEX				

TD952	C952YAH	Ford Transit 190D	Carlyle	B16F	1985	
TD957	C957YAH	Ford Transit 190D	Carlyle	B16F	1985	
TD958	C958YAH	Ford Transit 190D	Carlyle	B16F	1985	
TD972	C972YAH	Ford Transit 190D	Dormobile	B16F	1985	
TD975	C975YAH	Ford Transit 190D	Dormobile	B16F	1985	
TD982	C982YAH	Ford Transit 190D	Dormobile	B16F	1985	
TD984	C984YAH	Ford Transit 190D	Dormobile	B16F	1985	
TD985	C725FKE	Ford Transit 190D	Dormobile	B16F	1986	Ex East Kent, 1989
TD986	C726FKE	Ford Transit 190D	Dormobile	B16F	1986	Ex East Kent, 1989
TD987	C727FKE	Ford Transit 190D	Dormobile	B16F	1986	Ex East Kent, 1989

Livery: Ivory, red and orange;

Previous Registrations:

6149KP	WWA300Y, 9258VC, GAC98Y	7694VC	FWH37Y
6920MX	RUP388M	URL856S	WRO447S, XRL965

EASTERN NATIONAL

Eastern National Ltd, Stapleford Close, New Writtle Street, Chelmsford CM2 0SD

Depots and outstations: Anchor Street, Bishops Stortford; Fairfield Road, Braintree; Duke Street, Chelmsford; Telford Road, Clacton-on-Sea; Haven Road, Colchester; Queen Street, Colchester; Dunmow; Station Road, Harwich; High Street, Maldon and Walton-on-the-Naze.

202-223

Mercedes-Benz L608D   Reeve Burgess   B20F   1985-86

202	C202HJN	204	C204HJN	206w	C206HJN	210	C210HJN	215	C215HJN
203	C203HJN	205	C205HJN	207w	C207HJN	212	C212HJN	217	C217HJN

228-236

Mercedes-Benz L608D   Reeve Burgess   B20F   1986   Ex City Line, 1993

228	C482BHY	230	C485BHY	232	C489BHY	234	C494BHY	236	C496BHY
229	C484BHY	231	C486BHY	233	C493BHY	235	C495BHY		

237-246

Mercedes-Benz L608D   Reeve Burgess   B20F*   1985-86 Ex Western National, 1993
*242-6 are B19F

237	C678ECV	239	C685ECV	241	C688ECV	243	C697ECV	245	C700ECV
238	C684ECV	240	C687ECV	242	C695ECV	244	C698ECV	246	C964GCV

247	D534KGL	Mercedes-Benz L608D	Robin Hood	B20F	1986	Ex Western National, 1994
248	C107HGL	Mercedes-Benz L608D	Reeve Burgess	B20F	1986	Ex Western National, 1994
249w	C231HCV	Mercedes-Benz L608D	Robin Hood	B20F	1986	Ex Western National, 1994
250	C232HCV	Mercedes-Benz L608D	Robin Hood	B20F	1986	Ex Western National, 1994
251	C990GCV	Mercedes-Benz L608D	Robin Hood	B20F	1986	Ex Western National, 1994
252	C230HCV	Mercedes-Benz L608D	Robin Hood	B20F	1986	Ex Western National, 1994

601-617

Mercedes-Benz 709D   Reeve Burgess Beaver   B23F   1991

601	H601OVW	605	H605OVW	609	H609OVW	612	J612UTW	615	J615UTW
602	H602OVW	606	H606OVW	610	J610UTW	613	J613UTW	616	J616UTW
603	H603OVW	607	H607OVW	611	J611UTW	614	J614UTW	617	J617UTW
604	H604OVW	608	H608OVW						

618-630

Mercedes-Benz 709D   Plaxton Beaver   B23F   1991

618	J618UTW	621	J621UTW	624	J624UTW	627	J627UTW	629	J629UTW
619	J619UTW	622	J622UTW	625	J625UTW	628	J628UTW	630	J630UTW
620	J620UTW	623	J623UTW	626	J626UTW				

Photographed in Fairfield Road, Braintree during August 1995 was recently delivered 663, M663VJN, a Mercedes-Benz 709D with Plaxton Beaver bodywork. This batch of twenty were numerically succeeded by the sole example of the type with Frontline of Tamworth, that small operation being sold by FirstBus during the year. *Colin Lloyd*

Photographed in Colchester is Eastern National 807, L807OPU one of the 1994 delivery of Plaxton Pointer-bodied Dennis Darts. Note that on later deliveries of this body the high side lights have been moved to the front of the vehicle either side of the destination box. *Paul Wigan*

631-656

Mercedes-Benz 709D   Plaxton Beaver   B23F   1993-94

631	K631GVX	637	K637GVX	642	K642GVX	647	L647MEV	652	L652MEV
632	K632GVX	638	K638GVX	643	K643GVX	648	L648MEV	653	L653MEV
633	K633GVX	639	K639GVX	644	K644GVX	649	L649MEV	654	L654MEV
634	K634GVX	640	K640GVX	645	K645GVX	650	L650MEV	655	L655MEV
635	K635GVX	641	K641GVX	646	K646GVX	651	L651MEV	656	L656MEV
636	K636GVX								

657-676

Mercedes-Benz 709D   Plaxton Beaver   B23F   1995

657	M657VJN	661	M661VJN	665	M665VJN	669	M669VJN	673	M673VJN
658	M658VJN	662	M662VJN	666	M166VJN	670	M670VJN	674	M674VJN
659	M659VJN	663	M663VJN	667	M667VJN	671	M671VJN	675	M675VJN
660	M660VJN	664	M664VJN	668	M668VJN	672	M672VJN	676	M676VJN

677	L21AHA	Mercedes-Benz 709D	Plaxton Beaver	B23F	1993	Ex Frontline, 1995

801-822

Dennis Dart 9SDL3034   Plaxton Pointer   B34F   1993-94

801	L801MEV	806	L806OPU	811	L811OPU	815	L815OPU	819	L819OPU
802	L802MEV	807	L807OPU	812	L812OPU	816	L816OPU	820	L820OPU
803	L803OPU	808	L808OPU	813	L813OPU	817	L817OPU	821	L821OPU
804	L804OPU	809	L809OPU	814	L814OPU	818	L818OPU	822	L822OPU
805	L805OPU	810	L810OPU						

823-830

Dennis Dart 9.8SDL3054   Plaxton Pointer   B39F   1995

823	N823APU	825	N825APU	827	N827APU	829	N829APU	830	N830APU
824	N824APU	826	N826APU	828	N828APU				

1001-1007

Leyland Tiger TRBTL11/2R   Duple Dominant   DP47F   1983-84 Ex Yorkshire Rider, 1995

1001	EWR651Y	1003	EWR653Y	1005	A663KUM	1006	A665KUM	1007	A668KUM
1002	EWR652Y	1004	A660KUM						

1121w	HHJ382Y	Leyland Tiger TRCTL11/3R	Alexander TE	C53F	1983	
1128	B696WAR	Leyland Tiger TRCTL11/3R	Plaxton Paramount 3500 II	C51F	1985	
1129	B697WAR·	Leyland Tiger TRCTL11/3R	Plaxton Paramount 3500 II	C51F	1985	
1130	C130HJN	Leyland Tiger TRCTL11/3R	Plaxton Paramount 3200 II E	C53F	1986	
1131	EWW946Y	Leyland Tiger TRCTL11/3R	Plaxton Paramount 3200 E	C53F	1983	Ex Yorkshire Rider, 1995

1401-1429

Leyland Lynx LX112L10ZR/1R   Leyland Lynx   B49F*   1988   1427-9 are B47F

1401	E401HWC	1407	F407LTW	1414	F414MNO	1425	F425MJN	1428	F428MJN
1402	F402LTW	**1408**	F408LTW	**1415**	F415MWC	**1426**	F426MJN	**1429**	F429MJN
1403	F403LTW	**1413**	F413MNO	**1416**	F416MWC	**1427**	F427MJN		

1832-1924

Leyland National 11351A/1R   B49F   1978-79

1832	VAR898S	1860w	YEV318S	1865	YEV323S	1874	BNO664T	1914w	JHJ140V
1833	VAR899S	**1861**	YEV319S	**1867**	YEV325S	**1885**	BNO675T	**1916**	JHJ142V
1844w	WJN564S	**1862**	YEV320S	**1870**	YEV328S	**1890**	BNO680T	**1921**	JHJ147V
1850	YEV308S	**1863**	YEV321S	**1872**	ANO271S	**1899**	DAR121T	**1924**	JHJ150V
1851	YEV309S								

1933	MHJ729V	Leyland National 2 NL116L11/1R		B49F	1980
1935w	MHJ731V	Leyland National 2 NL116L11/1R		B49F	1980
2383	WNO479	Bristol KSW5G	Eastern Coach Works	O33/28R	1953
2384	WNO480	Bristol KSW5G	Eastern Coach Works	O33/28R	1953

3069-3094

Bristol VRT/SL3/6LXB   Eastern Coach Works   H39/31F   1980-81

3069	KOO787V	3076	KOO794V	3079	STW23W	3084	STW28W	3093	STW37W
3071	KOO789V	**3077**	STW21W	**3083**	STW27W	**3092**	STW36W	**3094**	STW38W
3072	KOO790V	**3078**	STW22W						

3103	UAR593W	Bristol VRT/SL3/6LXB	Eastern Coach Works	H43/31F	1981	
3106	UAR596W	Bristol VRT/SL3/6LXB	Eastern Coach Works	H43/34F	1981	
3109	UAR599W	Bristol VRT/SL3/6LXB	Eastern Coach Works	H43/31F	1981	
3112	XHK217X	Bristol VRT/SL3/6LXB	Eastern Coach Works	H43/34F	1981	
3127	XHK232X	Bristol VRT/SL3/6LXB	Eastern Coach Works	H43/31F	1981	
3219	VTH941T	Bristol VRT/SL3/501	Eastern Coach Works	H43/31F	1978	Ex Brewers, 1990
3220	WTH949T	Bristol VRT/SL3/501	Eastern Coach Works	H43/31F	1979	Ex Brewers, 1990
3221	WTH958T	Bristol VRT/SL3/501	Eastern Coach Works	H43/31F	1979	Ex Brewers, 1990
3222	BEP963V	Bristol VRT/SL3/501	Eastern Coach Works	H43/31F	1980	Ex Brewers, 1990
3223	MFA721V	Bristol VRT/SL3/501	Eastern Coach Works	DPH39/28F	1980	Ex PMT, 1994

3224-3233

Bristol VRT/SL3/6LXB   Eastern Coach Works   H43/31F   1977-81 Ex Yorkshire Rider, 1994-95
3226 was rebodied 1979

3224	AYG848S	3226	DWU298T	3228	LWU469V	3230	SUB789W	3232	JWT760V
3225	AYG850S	**3227**	LUA716V	**3229**	PWY44W	**3231**	SWW302R	**3233**	LUA717V

3500	WNO546L	Bristol VRT/SL2/6LX	Eastern Coach Works	O39/31F	1973
3501	NPU974M	Bristol VRT/SL2/6LX	Eastern Coach Works	O39/31F	1973

4007-4021

Leyland Olympian ONLXB/1R   Eastern Coach Works   DPH42/30F   1986   4010/12 ex Thamesway, 1995/92

4007	C407HJN	4012	C412HJN	4015	C415HJN	4017	C417HJN	4019	C419HJN
4008	C408HJN	**4013**	C413HJN	**4016**	C416HJN	**4018**	C418HJN	**4021**	C421HJN
4010	C410HJN	**4014**	C414HJN						

4501	B689BPU	Leyland Olympian ONTL11/2RHSp	Eastern Coach Works	CH45/28F	1985
4503	B691BPU	Leyland Olympian ONTL11/2RHSp	Eastern Coach Works	CH45/28F	1985
4510	D510PPU	Leyland Olympian ONTL11/2RHSp	Eastern Coach Works	CH45/28F	1986
4511	D511PPU	Leyland Olympian ONTL11/2RHSp	Eastern Coach Works	CH45/24F	1986
4512	D512PPU	Leyland Olympian ONTL11/2RHSp	Eastern Coach Works	CH45/24F	1986

Livery: Green and yellow.

Opposite: **Two vehicles representing the Eastern National fleet are** *(top)* **Leyland National 1933, MHJ729V, the sole remaining operational National 2 in the fleet and** *(bottom)* **4512, D512PPU, one of five Leyland Olympians with the final version of the Eastern Coach Works body design.** *Paul Wigan/Gerald Mead*

GRAMPIAN

Grampian Regional Transport Ltd, 395 King Street, Aberdeen AB9 1SP

1	K1GRT	Mercedes-Benz 0405G	Alexander Cityranger	AB60T	1992	
23	YSO231T	Leyland Atlantean AN68A/1R	Alexander AL	O45/29D	1978	
24	GRS114E	Leyland Atlantean PDR1/1	Alexander A	O43/34F	1967	
25	CRG325C	Daimler CVG6	Alexander B	H37/29R	1965	
31	LSK570	MCW MetroRider MF154/26	MCW	C29F	1989	Ex Mair, Dyce, 1992
40	ESK955	Volvo B58-56	Duple Dominant II	C51F	1979	Ex Kirkpatrick, Banchory, 1991
41	ESK956	Volvo B10M-61	Plaxton Paramount 3200 III	C53F	1988	Ex Alexander's, Aberdeen, 1989
44	ESK957	Volvo B10M-61	Plaxton Paramount 3200 III	C57F	1988	Ex Alexander's, Aberdeen, 1989
47	LSK476	Dennis Javelin 8.5SDL1903	Duple 320	C35F	1989	Ex Dewar, Falkirk, 1992
49	LSK475	Leyland Tiger TRCL10/3ARZM	Plaxton 321	C53F	1990	Ex Dewar, Falkirk, 1992
52	PSU623	Leyland Tiger TRCLXC/3RH	Plaxton Paramount 3200 E	C53F	1986	
53	PSU624	Leyland Tiger TRCLXC/3RH	Plaxton Paramount 3200 E	C53F	1986	
87	TSU651	Volvo B10M-60	Jonckheere Deauville P599	C51FT	1989	
88	TRS333	Volvo B10M-61	Jonckheere Jubilee P50	C46FT	1987	Ex Hallmark, Luton, 1990
89	PSU968	Volvo B10M-61	Jonckheere Jubilee P599	C51FT	1987	Ex Tellings-Golden Miller, Byfleet, 1990
92	J11GRT	Volvo B10M-60	Jonckheere Deauville P599	C49FT	1992	
93	K3GRT	Volvo B10M-60	Jonckheere Deauville P599	C49FT	1993	
94	K4GRT	Volvo B10M-60	Jonckheere Deauville P599	C47FT	1993	

101-110		Leyland Olympian ONLXB/1R	Alexander RH	H47/26D*	1984	*101-3 are H47/24D 101-3/8 are ONLXB/1R(6LXCT)

101	A101FSA	103	A103FSA	105	A105FSA	107	A107FSA	109	A109FSA
102	A102FSA	104	A104FSA	106	A106FSA	108	A108FSA	110	A110FSA

The GRT fleet still use dual-doored buses on city services. Shown here is 117, B117MSO, a Leyland Olympian with Alexander RH-type bodywork and one of five of this batch to be re-powered with Gardner 6LXCT units. *Reg Wilson*

112-121

Leyland Olympian ONLXB/1RV*   Alexander RH   H47/26D*   1985   *121 is H47/24D
*113/7/8/20/1 are ONLXB/1RV(6LXCT)

| 112 | B112MSO | 114 | B114MSO | 116 | B116MSO | 118 | B118MSO | 120 | B120MSO |
| 113 | B113MSO | 115 | B115MSO | 117 | B117MSO | 119 | B119MSO | 121 | B121MSO |

| 122 | E122DRS | Leyland Olympian ONCL10/2RZ | Alexander RH | DPH47/33F | 1988 |
| 123 | E123DRS | Leyland Olympian ONCL10/2RZ | Alexander RH | DPH47/33F | 1988 |

124-131

Leyland Olympian ONCL10/2RZ   Alexander RH   H49/29D   1988

| 124 | E124DRS | 126 | E126DRS | 128 | E128DRS | 130 | E130DRS | 131 | E131DRS |
| 125 | E125DRS | 127 | E127DRS | 129 | E129DRS | | | | |

228-237

Leyland Atlantean AN68A/1R   Alexander AL   H45/29D   1978

| 228 | YSO228T | 234 | YSO234T | 235 | YSO235T | 236 | YSO236T | 237 | YSO237T |

238-257

Leyland Atlantean AN68A/1R   Alexander AL   H45/29D   1979

238	DSA238T	242	DSA242T	246	DSA246T	250	DSA250T	254	DSA254T
239	DSA239T	243	DSA243T	247	DSA247T	251	DSA251T	256	DSA256T
240	DSA240T	244	DSA244T	248	DSA248T	252	DSA252T	257	DSA257T
241	DSA241T	245	DSA245T	249	DSA249T	253	DSA253T		

261-300

Leyland Atlantean AN68A/1R   Alexander AL   H45/29D   1980

261	HRS261V	269	HRS269V	277	HRS277V	285	HSO285V	293	LRS293W
262	HRS262V	270	HRS270V	278	HRS278V	286	HSO286V	294	LRS294W
263	HRS263V	271	HRS271V	279	HRS279V	287	HSO287V	295	LRS295W
264	HRS264V	272	HRS272V	280	HRS280V	288	HSO288V	296	LRS296W
265	HRS265V	273	HRS273V	281	HSO281V	289	HSO289V	297	LRS297W
266	HRS266V	274	HRS274V	282	HSO282V	290	HSO290V	298	LRS298W
267	HRS267V	275	HRS275V	283	HSO283V	291	LRS291W	299	LRS299W
268	HRS268V	276	HRS276V	284	HSO284V	292	LRS292W	300	LRS300W

301-315

Leyland Atlantean AN68C/1R   Alexander AL   H45/29D   1981

301	NRS301W	304	NRS304W	307	NRS307W	310	NRS310W	313	NRS313W
302	NRS302W	305	NRS305W	308	NRS308W	311	NRS311W	314	NRS314W
303	NRS303W	306	NRS306W	309	NRS309W	312	NRS312W	315	NRS315W

316-330

Leyland Atlantean AN68C/1R   Alexander AL   H45/29D   1982

316	URS316X	319	URS319X	322	URS322X	325	URS325X	328	URS328X
317	URS317X	320	URS320X	323	URS323X	326	URS326X	329	URS329X
318	URS318X	321	URS321X	324	URS324X	327	URS327X	330	URS330X

331-345

Leyland Atlantean AN68D/1R   Alexander AL   H45/29D   1983

331	XSS331Y	334	XSS334Y	337	XSS337Y	340	XSS340Y	343	XSS343Y
332	XSS332Y	335	XSS335Y	338	XSS338Y	341	XSS341Y	344	XSS344Y
333	XSS333Y	336	XSS336Y	339	XSS339Y	342	XSS342Y	345	XSS345Y

401-409

Mercedes-Benz 709D   Alexander AM   B23F   1993

| 401 | K401HRS | 403 | K403HRS | 405 | K405HRS | 407 | K407HRS | 409 | K409HRS |
| 402 | K402HRS | 404 | K404HRS | 406 | K406HRS | 408 | K408HRS | | |

| 432 | 2GRT | MCW MetroRider MF150/10 | MCW | C23F | 1987 |
| 433 | D33XSS | MCW MetroRider MF150/10 | MCW | C23F | 1987 |

434-439

Mercedes-Benz 709D   Reeve Burgess Beaver   B23F   1991

| 434 | H34USO | 436 | H36USO | 437 | H37USO | 438 | H38USO | 439 | H39USO |
| 435 | H35USO | | | | | | | | |

Shown in Grampian Gold Service livery is 519, M519RSS. This Mercedes-Benz O405 has Optare Prisma bodywork and the attractive livery blends well with this modern vehicle's low-floor profile.
Phillip Stephenson

440	E106JNH	Renault-Dodge S56	Alexander AM	DP23F	1987	Ex Northampton, 1994
441	E110JNH	Renault-Dodge S56	Alexander AM	DP23F	1988	Ex Northampton, 1994
442	E108JNH	Renault-Dodge S56	Alexander AM	DP23F	1988	Ex Mair's Coaches, 1994

501-514

Mercedes-Benz O405 — Wright Cityranger — B49F — 1993

501	L501KSA	504	L504KSA	507	L507KSA	510	L510KSA	513	L513KSA
502	L502KSA	505	L505KSA	508	L508KSA	511	L511KSA	514	L514KSA
503	L503KSA	506	L506KSA	509	L509KSA	512	L512KSA		

515-524

Mercedes-Benz O405 — Optare Prisma — B49F — 1995

| 515 | M1GRT | 517 | M517RSS | 519 | M519RSS | 521 | M521RSS | 523 | M523RSS |
| 516 | M516RSS | 518 | M518RSS | 520 | M520RSS | 522 | M522RSS | 524 | M524RSS |

Previous Registrations:

2GRT	D32XSS			
ESK955	HSE696V	LSK570	F632JSA	
ESK956	F101HSO	PSU623	D52VSO	
ESK957	F104HSO	PSU624	D53VSO	
LSK475	H154DJU	PSU968	D318VVV	
LSK476	G166HMS	TRS333	D330VVV	
		TSU651	F87CBD	

Livery: Cream and green; green (Aberdeen Corporation) 25

Opposite: **The typical colour of the stone buildings in the Granite City provides the background for these two pictures taken in Aberdeen. The 1991 delivery of Mercedes-Benz 709s carry Reeve Burgess bodywork as illustrated in the upper picture by 439, H39USO. Of interest is the Beaver Bus marketing name employed. The lower picture shows 507, L507KSA. This vehicle is also a Mercedes-Benz — an O405 unit with Wright Cityranger bodywork. This air-conditioned vehicle is painted in City Quick colours for a park and ride service.** *Tony Wilson*

KINGS of Dunblane

Kings of Dunblane Ltd, Carmuirs House, 300 Stirling Road, Larbert FK5 3NJ

KD1	ANK316X	Leyland Tiger TRCTL11/3R	Plaxton Supreme IV	C57F	1982	
KD2	FNM868Y	Leyland Tiger TRCTL11/2R	Plaxton Supreme V	C53F	1983	
KD3	D591MVR	Leyland Tiger TRCTL11/3RZ	Plaxton Paramount 3200 III	C53F	1987	Ex Rover, Horsley, 1993
KD4	D599MVR	Leyland Tiger TRCTL11/3RZ	Plaxton Paramount 3200 III	C53F	1987	Ex Shearings, 1992
KD5	E60MMT	Leyland Tiger TRCTL11/3RZ	Duple 340	C55F	1987	
KD6	F716SML	Leyland Tiger TRCL10/3ARZA	Duple 340	C55F	1989	
KD7	H838SLS	Leyland Tiger TRCL10/3ARZA	Plaxton Paramount 3500 III	C53F	1991	
KD8	J310XLS	Mercedes-Benz 811D	Reeve Burgess Beaver	C25F	1992	

Livery: White and blue

Kings of Dunblane Limited joined GRT holdings plc shortly before the merger with Badgerline. While remaining a separate operating unit the company is now based at Midland Bluebird's Larbert depot, between Falkirk and Denny. One of two Cummins-engined Leyland Tigers KD7, H838SLS, has Plaxton Paramount 3500 coachwork. *Kings of Dunblane*

KIRKPATRICK

Kirkpatrick of Deeside Ltd, 395 King Street, Aberdeen AB9 1SP

Depot : Dee Street, Banchory

903	PSU630	Leyland Leopard PSU5C/4R	Plaxton Supreme IV	C57F	1979	Ex Mair's Coaches, 1991
904	LSK546	Mercedes-Benz L608D	Plaxton Mini Supreme	C20F	1985	Ex Epsom Coaches, 1990
905	781GRT	Leyland Tiger TRCLXC/2RH	Plaxton Paramount 3200 E	C53F	1986	Ex Grampian, 1991
906	LSK527	Dennis Javelin 8.5SDL1903	Duple 320	C35F	1988	Ex Dewar, Falkirk, 1992
907	C103KDS	Mercedes-Benz L608D	Alexander AM	B21F	1986	Ex Midland Bluebird, 1993
908	C805SDY	Mercedes-Benz L608D	Alexander AM	B20F	1986	Ex Midland Bluebird, 1993
909	D951VSS	Mercedes-Benz 307D	Devon Conversions	M12	1986	Ex Rigblast, Dyce, 1993
910	E467JSG	Renault-Dodge S56	Alexander AM	B25F	1987	Ex SMT, 1994
911	YJF17T	Leyland Tiger TRCTL11/2R	Plaxton Supreme V Express	C53F	1982	Ex Leicester Citybus, 1995
912	LSK573	Mercedes-Benz 609D	Scott	C24F	1988	Ex Mair's Coaches, 1995
913	HSU955	Leyland Leopard PSU3F/4R	Plaxton Supreme IV Express	C48F	1980	Ex Mair's Coaches, 1995

Previous Registrations:

781GRT	D55VSO		LSK546	C200HGF	PSU630	FDF272T
HSU955	MAP346W		LSK573	F327WCS	WSU447	FDF275T
LSK527	E151XHS					

Livery: Cream and red

The Plaxton Paramount 3200 Express body style is shown here on Kirkpatrick's 905, 781GRT. This 1986 Leyland Tiger is one of the more rare Gardner-engined versions and was photographed outside its former home, the King Street depot in Aberdeen, having been new to Grampian. Kirkpatrick of Deeside is now controlled from Aberdeen but currently retains its Banchory depot. *Les Peters*

LEICESTER CITYBUS

Leicester Citybus Ltd, Abbey Park Road, Leicester, LE4 5AH

7	XDU178	Volvo B10M-61	Van Hool Alizée	C51FT	1985	Ex Mair's Coaches, 1995
8	542GRT	Volvo B10M-61	Jonckheere Jubilee P599	C51FT	1987	Ex The Londoners, 1995
18	BUT18Y	Leyland Tiger TRCTL11/2R	Plaxton Paramount 3200 E	C49F	1983	
19	FFK312	Leyland Tiger TRCTL11/2R	Plaxton Paramount 3200 E	C49F	1983	
23	B160WRN	Leyland Tiger TRCTL11/3RH	Duple Laser 2	C53F	1985	Ex Ribble, 1988
24	B165WRN	Leyland Tiger TRCTL11/3RH	Duple Laser 2	C53F	1985	Ex Ribble, 1988
25	GMS297S	Leyland Leopard PSU3E/4R	Alexander AYS	B53F	1978	Ex Midland Bluebird, 1994
26	ULS328T	Leyland Leopard PSU3E/4R	Alexander AYS	B53F	1979	Ex Midland Bluebird, 1994
27	GMS302S	Leyland Leopard PSU3D/4R	Alexander AYS	B53F	1978	Ex Grampian, 1995

31-35

MCW Metrobus DR102/35 — Alexander RL — H45/33F — 1983

31	AUT31Y	32	AUT32Y	33	AUT33Y	34	AUT34Y	35	AUT35Y

40-56

Dennis Dominator DDA142* — East Lancashire — H43/33F — 1981-82 *49-52 DDA141, 53-56 DDA146

40	TBC40X	44	TBC44X	48	TBC48X	51	TBC51X	54	TBC54X
41	TBC41X	45	TBC45X	49	TBC49X	52	TBC52X	55	TBC55X
42	TBC42X	46	TBC46X	50	TBC50X	53	TBC53X	56	TBC56X
43	TBC43X	47	TBC47X						

57-78

Dennis Dominator DDA155* — East Lancashire — H43/33F — 1982-83 *70 is DDA160, 71-4 are DDA173, 75-8 are DDA168

57	VAY57X	62	XJF62Y	67	XJF67Y	71	A71FRY	75	A75FRY
58	VAY58X	63	XJF63Y	68	XJF68Y	72	A72FRY	76	A76FRY
59	VAY59X	64	XJF64Y	69	XJF69Y	73	A73FRY	77	A77FRY
60	XJF60Y	65	XJF65Y	70	AUT70Y	74	A74FRY	78	A78FRY
61	XJF61Y	66	XJF66Y						

79-86

Dennis Dominator DDA1102* — East Lancashire — H43/33F — 1984-85 *81-3 are DDA1002 *84-86 are DDA901

79	B79MJF	81	B81MJF	83	B83MJF	85	B85MRY	86	B86MRY
80	B80MJF	82	B82MJF	84	B84MRY				

87-99

Dennis Dominator DDA1015 — East Lancashire — H46/33F — 1988

87	E87HNR	90	E90HNR	93	E93HNR	96	E96HNR	98	E98HNR
88	E88HNR	91	E91HNR	94	E94HNR	97	E97HNR	99	E99HNR
89	E89HNR	92	E92HNR	95	E95HNR				

The changes in city bus philosophy cannot be better demonstrated than by these three scenes. *Left.* The old order represented by Dennis Dominator 66, XJF66Y. *Opposite.* Modern single decker intake contrasts the Mercedes-Benz O405 501, M501GRY (top) with Dennis Falcon 623, L623XFP featuring Northern Counties Paladin bodywork. *Tony Wilson(2)/David Cole*

140-152

Dennis Dominator DDA1024 — East Lancashire — H46/33F — 1989

140	F140MBC	143	F143MBC	146	F146MBC	149	F149MBC	151	F151MBC
141	F141MBC	144	F144MBC	147	F147MBC	150	F150MBC	152	F152MBC
142	F142MBC	145	F145MBC	148	F148MBC				

154	FJF193	Leyland Titan PD2/1	Leyland	H33/29R	1950

179-200

Dennis Dominator DDA120* — East Lancashire — H43/33F — 1978-80 *188/98 are DDA110 200 is DDA110A

179	FUT179V	182	FUT182V	185w	FUT185V	187	FUT187V	198	YRY198T
180	FUT180V	183	FUT183V	186w	FUT186V	188	YRY188T	200	YRY200T
181	FUT181V	184	FUT184V						

205	NFP205W	Dennis Dominator DDA131	East Lancashire	H43/33F	1980
206	MUT206W	Dennis Dominator DDA131	East Lancashire	H43/33F	1980
226w	MUT226W	Dennis Dominator DDA120	Marshall	H43/33F	1980
229w	MUT229W	Dennis Dominator DDA120	Marshall	H43/33F	1980

239-264

Dennis Dominator DDA120* — East Lancashire — H43/33F — 1978-81 *239 is DDA101

239	UFP239S	245	FUT245V	252	MUT252W	256w	MUT256W	261	MUT261W
240	FUT240V	247w	FUT247V	253	MUT253W	257	MUT257W	262	MUT262W
241w	FUT241V	250	FUT250V	254	MUT254W	259	MUT259W	263	MUT263W
244	FUT244V	251	MUT251W	255	MUT255W	260w	MUT260W	264	MUT264W

265	ULS637X	MCW Metrobus DR102/28	Alexander RL	H45/33F	1982	Ex Midland Bluebird, 1994
266	ULS642X	MCW Metrobus DR104/10	Alexander RL	H45/33F	1982	Ex Midland Bluebird, 1994
267	BLS423Y	MCW Metrobus DR102/33	Alexander RL	H45/33F	1983	Ex Midland Bluebird, 1994
268	BLS432Y	MCW Metrobus DR102/33	Alexander RL	H45/33F	1983	Ex Midland Bluebird, 1994
269	BLS443Y	MCW Metrobus DR102/33	Alexander RL	H45/33F	1983	Ex Midland Bluebird, 1994
270	ULS636X	MCW Metrobus DR102/28	Alexander RL	H45/33F	1982	Ex Midland Bluebird, 1994

501-510

Mercedes-Benz O405 — Optare Prisma — B49F — 1995

501	M501GRY	503	M503GRY	505	M505GRY	507	M507GRY	509	M509GRY
502	M502GRY	504	M504GRY	506	M506GRY	508	M508GRY	510	M510GRY

611-619

Dennis Falcon SDA422 — East Lancashire 2000 — B48F — 1991-92

611	H611EJF	613	H613EJF	615	H615EJF	617	K617SBC	619	K619SBC
612	H612EJF	614	H614EJF	616	H616EJF	618	K618SBC		

620-626

Dennis Falcon SDA422 — Northern Counties Paladin — B48F — 1993

620	K620SBC	622	K622SBC	624	L624XFP	625	L625XFP	626	L626XFP
621	K621SBC	623	L623XFP						

718-737

Iveco Daily 49-10 — Carlyle Dailybus — B25F — 1989

718	F718PFP	724	F724PFP	728	G728WJU	732	G732WJU	735	G735WJU
720	F720PFP	725	F725PFP	729	G729WJU	733	G733WJU	736	G736WJU
721	F721PFP	726	F726PFP	730	G730WJU	734	G734WJU	737	G737WJU
723	F723PFP	727	G727WJU	731	G731WJU				

746-761

Renault-Dodge S56 — Northern Counties — B25F* — 1990-93 *753/4 are DP25F

746	K746VJU	750	K750VJU	753	J753MFP	756	J756MFP	759	J759NNR
748	K748VJU	751	H751ENR	754	J754MFP	757	J757MFP	760	K760SBC
749	K749VJU	752	H752ENR	755	J755MFP	758	J758NNR	761	K761SBC

763w	D321REF	Renault-Dodge S56	Northern Counties	B19F	1986	Ex Cleveland Transit, 1992
909	ARY225K	Scania BR111MH	MCW	B46D	1972	
911	D890MDB	Renault-Dodge S56	Northern Counties	B19F	1987	Ex Acorn, Chester, 1993
953	UFP175S	Scania BR111DH	MCW	H44/21F	1977	
954	UFP233S	Dennis Dominator DDA101	East Lancashire	H43/33F	1978	

Livery: Cream, red and maroon

Previous Registrations:

542GRT	E219GNV	FFK312	BUT19Y	XDU178	From new

In contrast to the 'slit' version of the East Lancashire body shown on page 43, Dennis 179, FUT179V displays the peaked-dome version similar to that introduced in 1963 on Leyland Atlanteans for Bolton. The picture was taken in Charles Street during June, 1995. *Gerald Mead*

As a change from Dennis Dominators, six Metrobuses joined Leicester Citybus during 1994 from Midland Bluebird — an indication of their origins is the typical Scottish Bus Group destination layout. Shown here is 266, ULS642X. *Tony Wilson*

LOWLAND

Lowland Omnibuses Ltd, Duke Street, Galashiels TD1 1QA

Depots : Duke Street, Galashiels; Dovecot Street, Hawick; Hospital Road, Haddington; Roxburgh Street, Kelso; Innerleithen Road, Peebles. **Outstations:** Tweedmouth Trading Estate, Berwick; Tantallon Road, North Berwick and Castlegate, Jedburgh.

2-39

						Seddon Pennine 7		Alexander AYS		DP49F*	1979-80 Ex Eastern Scottish 1985
											*5/32/7/9 are B53F, 7/10/31 are B60F

2	SSX602V	6	SSX606V	10	SSX610V	32	YSG632W	37	LSC937T
3	SSX603V	7	SSX607V	31	YSG631W	36	LSC936T	39	YSG639W
5	SSX605V								

43-59

				Seddon Pennine 7		Alexander AY		B60F	1977-78 Ex Eastern Scottish, 1985/87
43	ESC843S	48	ESC848S	56	VSX756R	58	VSX758R	59	VSX759R
46	ESC846S	49	ESC849S						

84	JFS984X	Seddon Pennine 7	Alexander AYS	B53F	1982	Ex Eastern Scottish, 1985
85	JFS985X	Seddon Pennine 7	Alexander AYS	DP49F	1982	Ex Eastern Scottish, 1985
86	JFS986X	Seddon Pennine 7	Alexander AYS	B53F	1982	Ex Eastern Scottish, 1985

101-105

				Scania K113CRB		Plaxton Paramount 3500 III	C53F	1990	Seating varies
101	G101RSH	102	LAT662	103	H103TSH	104	PSU314	105	PSU315

106	M106PKS	Scania K113CRB	Van Hool Alizée	C49FT	1995	
151	M151PKS	Scania N113CRL	Wright Axcess	DP49F	1995	
152	M152PKS	Scania N113CRL	Wright Axcess	DP49F	1995	
165	BSF765S	Leyland National 11351A/1R		B52F	1977	Ex Eastern Scottish, 1985
169	GSX869T	Leyland National 11351A/1R		B52F	1978	Ex Eastern Scottish, 1985

228-299

				Seddon Pennine 7		Alexander AT		C49F	1978-79 Ex Eastern Scottish, 1985
						258/65/7/8 ex SBG Engineering, 1990			

228	JSF928T	258	DSD958V	267	DSD967V	291	GSX891T	299	GSX899T
229	JSF929T	265	DSD965V	268w	DSD968V	292	GSX892T		

301	J301ASH	Leyland Tiger TR2R56V16Z4	Alexander Q	DP49F	1991	
302	J302ASH	Leyland Tiger TR2R56V16Z4	Alexander Q	DP49F	1991	
303	J303ASH	Leyland Tiger TR2R56V16Z4	Alexander Q	DP49F	1991	
304	J304ASH	Leyland Tiger TR2R56V16Z4	Alexander Q	DP49F	1991	
313	PSF313Y	Leyland Tiger TRBTL11/2R	Alexander AT	C49F	1982	Ex Eastern Scottish, 1985
314	PSF314Y	Leyland Tiger TRBTL11/2R	Alexander AT	C49F	1982	Ex Eastern Scottish, 1985
315	PSF315Y	Leyland Tiger TRBTL11/2R	Alexander AT	C49F	1982	Ex Eastern Scottish, 1985
316	PSF316Y	Leyland Tiger TRBTL11/2R	Alexander AT	C49F	1982	Ex Eastern Scottish, 1985

322-327

				Leyland Tiger TRBTL11/2RP	Alexander TE		C49F	1983	Ex Eastern Scottish, 1985
322	A322BSC	324	A324BSC	325	A325BSC	326	A326BSC	327	A327BSC
323	A323BSC								

Lowland was the first of the former Scottish Bus Group companies to be privatised, since then it has acquired the operations of Ian Glass in 1991 and Edinburgh Transit in 1994. New vehicles for the fleet have seen Scania coaches and, more recently, two Scania buses for the long established rail-replacement service in the Borders. Shown on the cover is 152, M152PKS, one of the first two Scanias in the UK meeting Euro2 emission requirements. Two vehicles from the Leyland range are seen opposite. *Above* is 328, D328DKS, one of three Tigers added to the fleet in 1987, while *below* is 960, B160KSC, an Olympian transferred from Eastern Scottish when the company was established in 1985. *Tony Wilson*

328	D328DKS	Leyland Tiger TRBTL11/2RH	Alexander TE	C49F	1987	
329	D329DKS	Leyland Tiger TRBTL11/2RH	Alexander TE	C49F	1987	
330	D330DKS	Leyland Tiger TRBTL11/2RH	Alexander TE	C49F	1987	
376	D276FAS	Leyland Tiger TRCTL11/3RH	Alexander TE	C53F	1987	Ex Highland Scottish, 1987
406	DFS806S	Seddon Pennine 7	Plaxton Supreme III Express	C49F	1978	Ex Eastern Scottish, 1985
407	DFS807S	Seddon Pennine 7	Plaxton Supreme III Express	C49F	1978	Ex Eastern Scottish, 1985
415	VFS324V	Seddon Pennine 7	Plaxton Supreme IV Express	C49F	1979	Ex Eastern Scottish, 1985
416	PSU316	Seddon Pennine 7	Plaxton Supreme IV Express	C45F	1980	Ex Eastern Scottish, 1985
417	PSU317	Seddon Pennine 7	Plaxton Supreme IV Express	C45F	1981	Ex Kelvin Scottish, 1986
418	PSU318	Seddon Pennine 7	Plaxton Supreme IV Express	C45F	1981	Ex Kelvin Scottish, 1986
433	PSU319	Seddon Pennine 7	Alexander AT	C45F	1979	Ex SBG Engineering, 1990
492	CSG792S	Seddon Pennine 7	Plaxton Supreme III Express	C49F	1978	Ex Eastern Scottish, 1985
494	CSG794S	Seddon Pennine 7	Plaxton Supreme III Express	C45F	1978	Ex Eastern Scottish, 1990
519	NSC413X	Leyland Tiger TRCTL11/3R	Duple Goldliner III	C46FT	1982	Ex Kelvin Scottish, 1988
520	PSU320	Leyland Royal Tiger B50	Roe Doyen	C44FT	1984	Ex Eastern Scottish, 1989
521	PSU321	Leyland Royal Tiger B50	Roe Doyen	C46FT	1984	Ex Eastern Scottish, 1988
522	KSU388	Leyland Tiger TRCTL11/3RH	Duple 340	C46FT	1987	Ex Kelvin Scottish, 1989
567	B267KPF	Leyland Tiger TRCTL11/2R	Plaxton Paramount 3200 II E	C53F	1985	Ex Ian Glass, Haddington, 1991
588	NSC411X	Leyland Tiger TRCTL11/3R	Duple Goldliner III	C50FT	1982	Ex Eastern Scottish, 1988
589	FFS9X	Leyland Tiger TRCTL11/3R	Duple Goldliner IV	C50FT	1982	Ex Eastern Scottish, 1988
590	KSU390	Leyland Tiger TRCTL11/2R	Plaxton Paramount 3200 E	C49F	1983	Ex Eastern Scottish, 1985
591	KSU391	Leyland Tiger TRCLXC/2RH	Plaxton Paramount 3200 E	C49F	1984	Ex Western Scottish, 1986
592	KSU392	Leyland Tiger TRCTL11/2RH	Plaxton Paramount 3200 II	C49F	1985	Ex Eastern Scottish, 1987
593	KSU393	Leyland Tiger TRCTL11/2RH	Plaxton Paramount 3200 II	C49F	1985	Ex Eastern Scottish, 1985
594	KSU394	Leyland Tiger TRCTL11/2RH	Plaxton Paramount 3200 III	C46FTL	1987	
604w	DSX400S	Bedford YMT	Plaxton Supreme III	C44DLT	1978	Ex Ian Glass, Haddington, 1991
606w	WOV582T	AEC Reliance 6U3ZR	Plaxton Supreme IV	C53F	1979	Ex Ian Glass, Haddington, 1991
613	PLS536W	Bedford YNT	Plaxton Supreme IV Express	C53F	1981	Ex Ian Glass, Haddington, 1991
615	KUX233W	Bedford YNT	Duple Dominant IV Express	C53F	1981	Ex Ian Glass, Haddington, 1991
616	FFS6X	Bedford YNT	Duple Dominant IV Express	C53F	1982	Ex Ian Glass, Haddington, 1991
617	FFS7X	Leyland Leopard PSU3E/4R	Plaxton Supreme IV	C53F	1982	Ex Ian Glass, Haddington, 1991
618	JFS166X	Leyland Tiger TRCTL11/3R	Plaxton Supreme V	C51F	1982	Ex Ian Glass, Haddington, 1991
621	ESX257	DAF MB200DKTL600	Van Hool Alizée	C46FT	1982	Ex Ian Glass, Haddington, 1991
622	BSS76	Leyland Tiger TRCTL11/3R	Jonckheere Jubilee	C51FT	1984	Ex Ian Glass, Haddington, 1991
623	KSU389	DAF SB2300DHS585	Smit Euroliner	C53F	1985	Ex Ian Glass, Haddington, 1991
626	C700USC	DAF MB200DKFL600	Duple Caribbean 2	C51F	1986	Ex Ian Glass, Haddington, 1991
627	KBZ3627	Dennis Javelin 12SDA1907	Duple 320	C53F	1988	Ex Ian Glass, Haddington, 1991
628	KBZ3628	Dennis Javelin 8.5SDL1903	Duple 320	C35F	1988	Ex Ian Glass, Haddington, 1991
629	KBZ3629	Dennis Javelin 12SDA1907	Duple 320	C53F	1989	Ex Ian Glass, Haddington, 1991
632	FFS10X	DAF MB200DKTL600	Plaxton Supreme VI	C57F	1982	Ex Shanks, Galashiels, 1992
655	TYS255W	Dennis Dominator DD135B	Alexander RL	H45/34F	1981	Ex Central Scottish, 1987
661	TYS261W	Dennis Dominator DD137B	Alexander RL	H45/34F	1981	Ex Central Scottish, 1987
664	TYS264W	Dennis Dominator DD137B	Alexander RL	H45/34F	1981	Ex Central Scottish, 1987
702	D402ASF	Renault-Dodge S56	Alexander AM	B21F	1986	Ex SMT, 1995
703	D403ASF	Renault-Dodge S56	Alexander AM	B21F	1986	Ex SMT, 1995
711w	D711CKS	Bedford VAS5	Reeve Burgess	DP27F	1987	
712w	D712CKS	Bedford VAS5	Reeve Burgess	DP25F	1987	
713w	D713CSH	Bedford VAS5	Reeve Burgess	B19F	1987	
714w	D714CSH	Bedford VAS5	Reeve Burgess	DP27F	1987	
715	G715OSH	Leyland Swift LBM6T/1RS	Reeve Burgess Harrier	C29F	1989	

720-727 — Optare MetroRider — Optare — DP32F* — 1994 — *720-2 are B32F

720	L720JKS	722	L722JKS	724	L724JKS	726	L726JKS	727	L727JKS
721	L721JKS	723	L723JKS	725	L725JKS	726	L726JKS		

733	E433JSG	Renault-Dodge S56	Alexander AM	B25F	1987	Ex SMT, 1995
734	E434JSG	Renault-Dodge S56	Alexander AM	B25F	1987	Ex SMT, 1995
748	G601OSH	Ford Transit VE6	Dormobile	C16F	1989	Ex Grieve, Hawick, 1994
749	H649USH	Ford Transit VE6	Deansgate	M15	1991	Ex Grieve, Hawick, 1994

750-756 — Renault-Dodge S56 — Alexander AM — B25F* — 1987-88 — *756 is DP25F

750	D750DSH	752	D752DSH	754	D754DSH	755	D755DSH	756	E756GSH
751	D751DSH	753	D753DSH						

771	H471OSC	Renault S75	Reeve Burgess Beaver	B31F	1991	Ex SMT, 1995
772	H472OSC	Renault S75	Reeve Burgess Beaver	B31F	1991	Ex SMT, 1995
776	H476OSC	Renault S75	Reeve Burgess Beaver	B31F	1991	Ex SMT, 1995
782	H502OSC	Renault S75	Reeve Burgess Beaver	B31F	1991	Ex SMT, 1995
850	MDL650R	Bristol VRT/SL3/6LXB	Eastern Coach Works	H43/31F	1976	Ex Southern Vectis, 1990
854	GSC854T	Leyland Fleetline FE30AGR	Eastern Coach Works	H43/32F	1978	Ex Eastern Scottish, 1985
855	NDL655R	Bristol VRT/SL3/6LXB	Eastern Coach Works	H43/31F	1977	Ex Southern Vectis, 1991
856	NDL656R	Bristol VRT/SL3/6LXB	Eastern Coach Works	H43/31F	1977	Ex Southern Vectis, 1991
864	OSG64V	Leyland Fleetline FE30AGR	Eastern Coach Works	H43/32F	1979	Ex SMT, 1995
871	OSG71V	Leyland Fleetline FE30AGR	Eastern Coach Works	H43/32F	1979	Ex Eastern Scottish, 1985
901	D901CSH	Leyland Olympian ONTL11/1RH	Alexander RL	DPH43/27F	1987	
902	D902CSH	Leyland Olympian ONTL11/1RH	Alexander RL	DPH43/27F	1987	
916	ALS116Y	Leyland Olympian ONLXB/1R	Alexander RL	H45/32F	1983	Ex Eastern Scottish, 1985
943	A143BSC	Leyland Olympian ONLXB/1R	Alexander RL	DPH45/29F	1984	Ex Eastern Scottish, 1985
959	B159KSC	Leyland Olympian ONTL11/1RH	Alexander RL	H45/32F	1985	Ex Eastern Scottish, 1985
960	B160KSC	Leyland Olympian ONTL11/1RH	Alexander RL	H45/32F	1985	Ex Eastern Scottish, 1986
988	VAO488Y	Leyland Titan TNTL11/1RF	Leyland	H48/31F	1982	Ex Ian Glass, Haddington, 1991
1014	RSX84J	Daimler Fleetline CRG6LXB	Alexander AD	O44/31F	1971	Ex Fife Scottish, 1986
1101	NUD801W	Leyland Leopard PSU3F/5R	Duple Dominant II	C53F	1981	Ex Lothian Transit, 1994
1108	D408ASF	Renault-Dodge S56	Alexander AM	B21F	1986	Ex SMT, 1995
1110	D410ASF	Renault-Dodge S56	Alexander AM	B21F	1986	Ex SMT, 1995
1122	PSU322	Seddon Pennine 7	Plaxton Supreme IV Express	C49F	1979	Ex Lothian Transit, 1994
1132	KEX532	Volvo B58-61	Plaxton Viewmaster IV Exp	C53F	1980	Ex Lothian Transit, 1994
1134	VXI8734	Leyland Leopard PSU3E/3R	Duple Dominant I	C49F	1977	Ex Lothian Transit, 1994
1135	E435JSG	Renault-Dodge S56	Alexander AM	B25F	1987	Ex SMT, 1995
1140	E440JSG	Renault-Dodge S56	Alexander AM	B25F	1987	Ex SMT, 1995
1151	OSG51V	Leyland Fleetline FE30AGR	Eastern Coach Works	H43/32F	1979	Ex SMT, 1995
1152	OSG52V	Leyland Fleetline FE30AGR	Eastern Coach Works	H43/32F	1979	Ex SMT, 1995
1153	OSG53V	Leyland Fleetline FE30AGR	Eastern Coach Works	H43/32F	1979	Ex SMT, 1995
1158	158ASV	Leyland National 11351A/1R		B52F	1978	Ex Lothian Transit, 1994
1159	D459CKV	Freight Rover Sherpa	Rootes	B16F	1986	Ex Lothian Transit, 1994
1161	BSF771S	Leyland National 11351A/1R		B51F	1977	Ex Lothian Transit, 1994
1168w	GSX868T	Leyland National 11351A/1R		B52F	1978	Ex Lothian Transit, 1994
1172	OSG72V	Leyland Fleetline FE30AGR	Eastern Coach Works	H43/32F	1979	Ex SMT, 1995
1175	K175YUE	Mercedes-Benz 811D	Wright Nim-bus	B33F	1992	Ex Lothian Transit, 1994

Previous Registrations:

158ASV	OLS805T	LAT662	G102RSH
BSS76	A678DSF	NSC411X	MSC557X, KSU388
ESX257	WFR612Y	NSC413X	MSC552X, WLT741, HGD741X, PSU319
FFS6X	PNT803X	NSC702X	FFS10X
FFS7X	MVK332X	PLS536W	ASH1W
FFS9X	PSF559Y, KSU389	PSU314	H104TSH
FFS10X	VTT14X, LAT662	PSU315	H105TSH
KBZ3627	E888MSX	PSU316	USX971V
KBZ3628	E900MSX	PSU317	DSC974W
KBZ3629	F777UFS	PSU318	DSC975W
KEX532	GOP730W, PBC453	PSU319	DSD933V
KSU388	D320RNS	PSU320	A562BSX
KSU389	B88KSF	PSU321	A563BSX
KSU390	TFS317Y	PSU322	LSC950T
KSU391	A185UGB	RSX84J	RXA51J, PSU314
KSU392	B342RLS	VFS324V	OSF963V, PSU315
KSU393	B343RLS	VXI8734	RRS49R
KSU394	D501CSH		

Livery: Green and yellow or cream, green and yellow: cream and turquoise (Ian Glass); red and cream (Lothian Transit) blue and yellow (Scottish Citylink) 101/2; metallic green (Prestige Tours) 104-6.

MAIR'S COACHES

G E Mair Hire Services Ltd, St Peter Street, Aberdeen AB2 3HU

Depot : King Street, Aberdeen

701	XWL539	Volvo B10M-61	Jonckheere Jubilee P599	C51FT	1987	Ex West Kingsdown Coach, 1989
702	PSU609	Volvo B10M-60	Plaxton Paramount 3500 III	C48FT	1989	Ex Wallace Arnold, 1992
703	PSU626	Volvo B10M-60	Plaxton Paramount 3500 III	C48FT	1989	Ex Clyde Coast, Ardrossan, 1992
704	PSU629	Volvo B10M-61	Van Hool Alizée	C53F	1984	
705	PSU628	Volvo B10M-61	Jonckheere Jubilee P599	C51FT	1987	Ex Buddens Skylark, Woodfalls, 1990
706	FSU333	Volvo B10M-60	Jonckheere Deauville P599	C51FT	1989	Ex Marbill, Beith, 1993
707	F326WCS	Mercedes-Benz 609D	Scott	C24F	1988	
708	F634JSO	Mercedes-Benz 609D	Made-to-Measure	C19F	1989	
709	F633JSO	Mercedes-Benz 609D	Made-to-Measure	C19F	1989	
710	K950HSA	Mercedes-Benz 609D	Made-to-Measure	C24F	1993	
711	FSU335	Mercedes-Benz L608D	Reeve Burgess	DP19F	1986	Ex Midland Bluebird, 1993
712	D532RCK	Mercedes-Benz L608D	Reeve Burgess	B20F	1986	Ex Midland Bluebird, 1993
713	LSK571	Mercedes-Benz L608D	Alexander	DP20F	1986	Ex Midland Bluebird, 1993
715	G143SUS	Mercedes-Benz 609D	Scott	C22F	1990	
716	H193CVU	Mercedes-Benz 609D	Made-to-Measure	C24F	1990	
717	D232UHC	Mercedes-Benz L608D	Alexander	DP20F	1986	Ex Bluebird Northern, 1992
718	JSV426	Leyland Tiger TRCTL11/3R	Plaxton Paramount 3500	C55F	1983	Ex Armchair, Brentford, 1990
720	PSU631	Volvo B10M-61	Jonckheere Jubilee P50	C53F	1985	Ex Buddens Skylark, Woodfalls, 1988
721	LSK572	Volvo B10M-61	Plaxton Paramount 3200 III	C57F	1988	Ex Grampian, 1991
722	WSU460	Volvo B10M-61	Van Hool Alizée	C53F	1985	Ex Selwyn, Runcorn, 1994
724	WSU480	Leyland Leopard PSU5C/4R	Plaxton Supreme IV	C53F	1978	Ex Grampian, 1992
725	WSU481	Leyland Leopard PSU3E/4RT	Plaxton Supreme IV Express	C48F	1980	Ex Grampian, 1991
726	LSU917	Leyland Leopard PSU3F/4R	Plaxton Supreme IV Express	C48F	1981	Ex Southdown, 1989
727	LSU717	Leyland Leopard PSU3F/4R	Plaxton Supreme IV Express	C48F	1981	Ex Southdown, 1989
728	HSO61N	Leyland Leopard PSU4C/4R	Alexander AY	C45F	1975	Ex Grampian, 1988
729	ORS60R	Leyland Leopard PSU4C/4R	Alexander AY	C45F	1977	Ex Grampian, 1993
730	YSO232T	Leyland Atlantean AN68A/1R	Alexander AL	H45/29D	1978	Ex Grampian, 1995
731	J11AFC	Volvo B10M-60	Jonckheere Deauville P599	C49FT	1992	
732	737ABD	Volvo B10M-61	Jonckheere Deauville P599	C51FT	1988	
733	PSU627	Volvo B10M-61	Jonckheere Deauville P599	C51FT	1989	Ex River Valley, Sutton Valence, 1991
734	WSU447	Volvo B10M-60	Jonckheere Deauville P599	C51FT	1990	Ex Redwing, Camberwell, 1994
736	131ASV	Volvo B10M-60	Van Hool Alizée	C55F	1990	Ex Henry Crawford, Neilston, 1993
737	K67HSA	Toyota Coaster HDB30R	Caetano Optimo II	C18F	1993	
738	L538XUT	Toyota Coaster HZB50R	Caetano Optimo III	C18F	1994	
739	LSK529	Dennis Javelin 8.5SDL1903	Plaxton Paramount 3200 III	C35F	1988	Ex Dewar, Falkirk, 1992
740	LSK530	Dennis Javelin 8.5SDL1903	Plaxton Paramount 3200 III	C35F	1988	Ex Dewar, Falkirk, 1992

Previous Registrations:

131ASV	G260RNS	PSU609	F408DUG
737ABD	F950RNV	PSU626	F986HGE
FSU333	G845GNV	PSU627	F913YNV
FSU335	D517RCK	PSU628	D95BNV
JSV426	FNM863Y	PSU629	B229LSO, 737ABD
K950HSA	K983XND	PSU631	B497CBD
LSK529	F739WMS	WSU447	G171RBD
LSK530	F369MUT	WSU460	B122DMA, SEL392
LSK571	D424UHC	WSU480	YYJ299T, 405DCD, FRX868T
LSK572	F103HSO	WSU481	GWV931V
LSU717	MAP349W	XWL539	D315VVV
LSU917	MAP348W		

Livery: Red and cream or red and gold; yellow and blue (Scottish Citylink) 701-3/5/6/36.

Mair's Coaches now share the King Street depot with Grampian. Here is seen 723, HSU955, a Leyland Leopard with Plaxton Supreme IV Express coachwork that was new to Southdown. It retains that operator's preferred 48-seat layout which avoided the bench seat for five at the rear. This vehicle was transferred to Kirkpatrick of Deeside in the early Autumn of 1995. *Les Peters*

Mair's operate contracts for which service vehicles are maintained. Seen in Union Street, Aberdeen is 728, HSO61N, a Leyland Leopard with Alexander Y-type body which has typified the Scottish bus scene since it first appeared in 1961. This model features a bus doorway and high-back seating and was transferred from the Grampian fleet in 1988. *Murdoch Currie*

MIDLAND BLUEBIRD

Midland Bluebird Ltd, Carmuirs House, 300 Stirling Road, Larbert FK5 3NJ

Depots: Dunmore Street, Balfron; Cowie Road, Bannockburn; Stirling Road, Larbert; High Street, Linlithgow.

26-47

| | | | | | | | | | Leyland National 2 NL116L11/1R | | | B52F* | 1980-81 41-7 ex Kelvin Scottish, 1987-88 *29 is B49F |

26	DMS26V	32	NLS984W	42	RFS587V	45	SNS830W	47	YFS303W
29	NLS981W	41	RFS580V	44	WAS764V	46	YFS302W		

51	K473EDT	Mercedes-Benz O405	Alexander Cityranger	B51F	1992	Ex Mercedes-Benz demonstrator, 1993

52-57

| | | | | Mercedes-Benz O405 | Wright Cityranger | B51F | 1993 | | |

52	L552GMS	54	L554GMS	55	L555GMS	56	L556GMS	57	L557GMS
53	L553GMS								

58	L140MAK	Mercedes-Benz O405	Wright Cityranger	B51F	1994	
101	FSU381	Leyland Tiger TRBTL11/2R	Alexander AT	C49F	1983	
102	FSU382	Leyland Tiger TRBTL11/2R	Alexander AT	C49F	1983	
103	FSU383	Leyland Tiger TRBTL11/2R	Alexander AT	C49F	1983	
105	FSU380	Leyland Tiger TRBTL11/2R	Alexander AT	C49F	1983	
107	101ASV	Leyland Tiger TRBTL11/2R	Duple Dominant II Express	C49F	1983	Ex Kelvin Scottish, 1986
108	FSU308	Leyland Tiger TRBTL11/2R	Duple Dominant II Express	C47F	1983	Ex Kelvin Scottish, 1986
114	FSU334	Leyland Tiger TRBTL11/2R	Alexander TE	C46F	1983	
116	SSU816	Leyland Tiger TRBTL11/2RP	Alexander TC	C47F	1983	
117	BSV807	Leyland Tiger TRBTL11/2RP	Alexander TC	C42DL	1983	Ex Kelvin Scottish, 1988
118	7881UA	Leyland Tiger TRBTL11/2RP	Alexander TE	C42DL	1983	
121	SSU821	Leyland Tiger TRCTL11/3RH	Duple Caribbean	C51F	1984	
124	VSU715	Leyland Tiger TRCTL11/3RH	Duple Caribbean	C44FT	1984	
129	SSU859	Leyland Tiger TRCTL11/3RH	Duple Laser	C46FT	1984	
130	693AFU	Leyland Tiger TRCTL11/3RH	Duple Laser	C46FT	1984	
131	SSU831	Leyland Tiger TRCTL11/3RH	Duple Laser 2	C46FT	1985	
132	SSU837	Leyland Tiger TRCTL11/3RH	Duple Laser 2	C46FT	1985	
136	SSU897	Leyland Tiger TRCLXC/2RH	Plaxton Paramount 3200 E	C49F	1984	Ex Clydeside Scottish, 1985
138	OVT798	Leyland Tiger TRCLXC/2RH	Plaxton Paramount 3200 E	C49F	1984	Ex Western Scottish, 1986
139	119ASV	Leyland Tiger TRCTL11/3R	Duple Goldliner IV	C53F	1982	Ex Western Scottish, 1986

141-145

| | | | | Leyland Tiger TRCTL11/3RH | Duple 340 | C49FT | 1987 | | |

141	SSU841	142	SSU857	143	KSU834	144	FSV634	145	156ASV

Photographed in Stirling during July 1995, while heading for Glasgow, was Midland Bluebird 154, PSU625. This Leyland Tiger is fitted with the express version of the Plaxton Paramount 3200 design and came from the Grampian fleet in 1991. The vehicle carries Bluebird Coaches fleetnames — not to be confused with Bluebird Buses the Stagecoach operation that also shares Alexander ancestry. *Les Peters*

147	SSU827	Leyland Tiger TRCTL11/3R	Duple Goldliner III	C53F	1982	Ex Kelvin Scottish, 1988
148	SSU861	Leyland Tiger TRCTL11/3R	Duple Goldliner III	C53F	1982	Ex Kelvin Scottish, 1988
149	SSU829	Leyland Tiger TRCTL11/3R	Duple Goldliner III	C53F	1982	Ex Kelvin Scottish, 1988
150	WSU487	Leyland Tiger TRCTL11/3R	Plaxton Paramount 3200 E	C53F	1983	Ex Grampian, 1991
151	WSU489	Leyland Tiger TRCTL11/3R	Plaxton Paramount 3200 E	C53F	1983	Ex Grampian, 1991
152	GSU338	Leyland Tiger TRCTL11/3R	Plaxton Paramount 3200 E	C57F	1983	Ex Grampian, 1991
153	GSU339	Leyland Tiger TRCTL11/3R	Plaxton Paramount 3200 E	C57F	1983	Ex Grampian, 1991
154	PSU625	Leyland Tiger TRCLXC/2RH	Plaxton Paramount 3200 E	C53F	1986	Ex Grampian, 1991
155	PSU622	Leyland Tiger TRCLXC/2RH	Plaxton Paramount 3200 E	C53F	1986	Ex Grampian, 1991
156	WSU479	Leyland Tiger TRCTL11/3RH	Plaxton Paramount 3200 E	C48FT	1984	Ex Grampian, 1993
157	YJF16Y	Leyland Tiger TRCTL11/2R	Plaxton Supreme V Express	C53F	1982	Ex Leicester Citybus, 1995
202	692FFC	Volvo B10M-61	Jonckheere Jubilee P599	C51FT	1989	Ex Laing, Thornton Heath, 1991
203	TSU682	Volvo B10M-61	Jonckheere Jubilee P599	C51FT	1989	Ex Grampian, 1992
204	ESK958	Volvo B10M-61	Plaxton Paramount 3200 III	C48FT	1989	Ex Grampian, 1992
205	FSU315	Volvo B10M-60	Plaxton Paramount 3200 III	C46FT	1989	Ex Wallace Arnold, 1993
207	144ASV	Volvo B10M-60	Jonckheere Deauville P599	C51FT	1990	Ex Redwing, Camberwell, 1994
208	K924RGE	Volvo B10M-60	Jonckheere Deauville P599	C51FT	1993	Ex Park's, 1995

327-356

Leyland Leopard PSU3E/4R — Alexander AYS — B53F — 1979

327	ULS327T	338	ULS338T	352	DLS352V	355	DLS355V	356	DLS356V
332	ULS332T	351	DLS351V	353	DLS353V				

360-366

Leyland Leopard PSU3E/4R — Alexander AT — C49F — 1980 — 366 ex Kelvin Scottish, 1985

360	EMS360V	362	EMS362V	363	EMS363V	364	EMS364V	366	EMS366V

374-386

Leyland Leopard PSU3F/4R — Alexander AYS — B53F* — 1980 — *378 is DP49F

374	LMS374W	377	LMS377W	379	LMS379W	382	LMS382W	386	LMS386W
376	LMS376W	378	LMS378W	381	LMS381W	384	LMS384W		

393	FSU318	Leyland Leopard PSU3G/4R	Duple Dominant II Express	C49F	1981
397	FSU302	Leyland Leopard PSU3G/4R	Duple Dominant II Express	C49F	1981
398	RMS398W	Leyland Leopard PSU3G/4R	Alexander AT	C49F	1981
399	RMS399W	Leyland Leopard PSU3G/4R	Alexander AT	C49F	1981
400	RMS400W	Leyland Leopard PSU3G/4R	Alexander AT	C49F	1981

403-412

Leyland Leopard PSU3G/4R — Alexander AYS — DP49F* — 1982 — *410/1 are B53F

403	TMS403X	408	TMS408X	410	TMS410X	411	TMS411X	412	TMS412X

413	ULS713X	Leyland Leopard PSU3G/4R	Alexander AT	C49F	1982
414	ULS714X	Leyland Leopard PSU3G/4R	Alexander AT	C49F	1982
416	ULS716X	Leyland Leopard PSU3G/4R	Alexander AT	C49F	1982
417	ULS717X	Leyland Leopard PSU3G/4R	Alexander AT	C49F	1982

The Alexander T-type replaced the Y-type after a period of concurrent production. An example of the later, all-aluminium AT variant is Midland Bluebird 417, ULS717X seen in Stirling on a Leyland Leopard chassis. The BN garage code is for Bannockburn, where the depot for Stirling is located.
Les Peters

No.	Reg.	Chassis	Body	Type	Year	Notes
419	WFS154W	Leyland Leopard PSU3F/4R	Alexander AYS	B53F	1980	Ex Alexander (Fife), 1982
421	XMS421Y	Leyland Leopard PSU3G/4R	Alexander AYS	DP49F	1982	
425	XMS425Y	Leyland Leopard PSU3G/4R	Alexander AYS	DP49F	1982	
451	GSO80V	Leyland Leopard PSU3E/4R	Alexander AYS	B53F	1980	Ex Fife Scottish, 1988
452	GSO81V	Leyland Leopard PSU3E/4R	Alexander AYS	B53F	1980	Ex Fife Scottish, 1988
454	WFS146W	Leyland Leopard PSU3F/4R	Alexander AYS	B53F	1980	Ex Fife Scottish, 1988
455	WFS145W	Leyland Leopard PSU3F/4R	Alexander AYS	B53F	1980	Ex Fife Scottish, 1988
456	WFS146W	Leyland Leopard PSU3F/4R	Alexander AYS	B53F	1980	Ex Fife Scottish, 1988
457	CFS155W	Leyland Leopard PSU3F/4R	Alexander AYS	B53F	1981	Ex Fife Scottish, 1988
467	HSU247	Leyland Leopard PSU5D/4R(TL11)	Plaxton Supreme IV	C53F	1981	Ex Grampian, 1991
468	HSU273	Leyland Leopard PSU5D/4R(TL11)	Plaxton Supreme IV	C53F	1981	Ex Grampian, 1991
501w	SSC108P	Seddon Pennine 7	Alexander AT	C24DL	1976	Ex Western Scottish, 1984
509	JSF909T	Seddon Pennine 7	Alexander AT	C49F	1978	Ex SMT, 1995
541	PXI8935	Seddon Pennine 7	Plaxton Supreme IV Express	C49F	1980	Ex Kelvin Central, 1989

551-566 Scania N113CRB   Wright Endurance   B49F   1994

551	L551HMS	555	L555HMS	558	L558JLS	561	L561JLS	564	L564JLS
552	L552HMS	556	L556HMS	559	L559JLS	562	L562JLS	565	L565JLS
553	L553HMS	557	L557JLS	560	L60HMS	563	L563JLS	566	L566JLS
554	L554HMS							

567-574 Scania N113CRL   Wright Axcess   B49F   1995

567	M567RMS	569	M569RMS	571	M571RMS	573	N573VMS	574	N574VMS
568	M568RMS	570	M570RMS	572	N572VMS				

604-609 Mercedes-Benz L608D   Alexander AM   B20F   1986   Ex Bluebird, 1992

604	C812SDY	606	C821SDY	607	D226UHC	608	D227UHC	609	D229UHC
605	C817SDY							

No.	Reg.	Chassis	Body	Type	Year	Notes
619	D119NUS	Mercedes-Benz L608D	Alexander AM	B21F	1986	Ex Kelvin Scottish, 1987
620	D120NUS	Mercedes-Benz L608D	Alexander AM	B21F	1986	Ex Kelvin Scottish, 1987
622	C102KDS	Mercedes-Benz L608D	Alexander AM	B21F	1986	Ex Kelvin Scottish, 1987
625	H925PMS	Mercedes-Benz 709D	Reeve Burgess Beaver	B25F	1990	
626	H926PMS	Mercedes-Benz 709D	Reeve Burgess Beaver	B25F	1990	

632-641 Mercedes-Benz 709D   Alexander AM   B25F*   1991   *637-641 are DP23F

632	H972RSG	634	H974RSG	636	H976RSG	638	J775WLS	640	J778WLS
633	H973RSG	635	H975RSG	637	J774WLS	639	J776WLS	641	J779WLS

651-659 Mercedes-Benz 709D   Alexander AM   B25F   1993

651	K651DLS	653	K653DLS	655	K655DLS	657	K657DLS	659	K659DLS
652	K652DLS	654	K654DLS	656	K656DLS	658	K658DLS		

No.	Reg.	Chassis	Body	Type	Year	Notes
662	L523KSX	Optare MetroRider	Optare	B31F	1994	Ex SMT, 1995
663	M284SMS	Optare MetroRider	Optare	DP25F	1995	Owned by Central RC

681-685 Renault S75   Reeve Burgess Beaver   B31F   1991   Ex SMT, 1994-95

681	H481OSC	682	H482OSC	683	H483OSC	684	H484OSC	685	H475OSC

702-730 Leyland Alteantean AN68A/1R   Alexander AL   H45/29D*   1976-78 Ex Grampian, 1991-95
*706/15/7/9/20/3 are H45/31F; 716 is O45/31F and is owned by Stirling District Council

702	ORS202R	708	ORS208R	716	ORS216R	720	XSA220S	724	XSA224S
703	ORS203R	709	ORS209R	717	ORS217R	721	XSA221S	725	XSA225S
704	ORS204R	710	ORS210R	718	XSA218S	722	XSA222S	727	XSA227S
705	ORS205R	711	ORS211R	718	XSA219S	723	XSA223S	730	YSO230T
706	ORS206R	715	ORS215R						

Opposite, top: **The latest arrivals with Midland Bluebird are Scania N113CRL chassis with Wright Axcess bodies. This low-floor chassis has become a popular product as the market for more accessible vehicles has developed rapidly during the last two years. Pictured in Edinburgh is 574, N574VMS.** *Opposite, bottom:* **Leyland Atlanteans have been cascaded from Grampian to several other GRT companies. Seen with Midland Bluebird is 721, XSA221S, which features the Alexander AL body in this case now converted to single-door.** *Tony Wilson/Phillip Stephenson*

In its later years, while part of the Scottish Bus Group, Midland Scottish's double-deck requirement was met by a large number of MCW Metrobuses. Unlike English operators who, in the main, used MCW for the bodywork, Midland mostly purchased Alexander products. The formation of Strathtay and Kelvin took large numbers of the type from the Midland fleet, now only 45 remain. Seen while operating a special service to the Highland games at Callander is 816, D116ELS. Note the additional 'window' on the lower deck. *R A Smith*

800-807				MCW Metrobus DR132/6		Alexander RL		H45/33F	1985	800-4 ex Kelvin Central, 1990
800	B100PKS	802	B102PKS	804	B104PKS	806	B106PKS	807	B88PKS	
801	B101PKS	803	B103PKS	805	B105PKS					

808-813				MCW Metrobus DR102/52		Alexander RL		H45/33F*	1986	*812/3 are DPH45/33F
808	D108ELS	810	D110ELS	811	D111ELS	812	143ASV	813	110ASV	
809	D109ELS									

814	HSU301	MCW Metrobus DR132/9	Alexander RL	DPH45/33F	1986
815	D115ELS	MCW Metrobus DR132/9	Alexander RL	DPH45/33F	1986
816	D116ELS	MCW Metrobus DR132/9	Alexander RL	DPH45/33F	1986

817-821				MCW Metrobus DR132/10		Alexander RL		DPH47/33F	1987	
817	365UMY	818	VXU444	819	WLT724	820	TSV612	821	373GRT	

830-834				MCW Metrobus DR102/28		Alexander RL		H45/33F	1982	831-4 ex Kelvin Central, 1989
830	ULS630X	831	ULS620X	832	ULS622X	833	ULS623X	834	ULS624X	

843	ULS643X	MCW Metrobus DR104/10	Alexander RL	H45/33F	1982
859	BLS437Y	MCW Metrobus DR102/33	Alexander RL	H45/33F	1983
868	BLS446Y	MCW Metrobus DR102/33	Alexander RL	H45/33F	1983
870	A470GMS	MCW Metrobus DR102/39	Alexander RL	H45/33F	1984
877	A477GMS	MCW Metrobus DR102/39	Alexander RL	H45/33F	1984
881	B581MLS	MCW Metrobus DR102/39	Alexander RL	H45/33F	1984
882	B582MLS	MCW Metrobus DR102/39	Alexander RL	H45/33F	1984
883	B583MLS	MCW Metrobus DR132/2	Alexander RL	H45/33F	1984

Five Renault-Dodge S75s with Reeve Burgess Beaver bodywork in the current fleet were transferred from SMT during 1994-95 and have been put into service from Linlithgow. Shown here is 685, H475OSC. *Phillip Stephenson*

884	B584MLS	MCW Metrobus DR132/2	Alexander RL	H45/33F	1984	Ex Kelvin Central, 1990
885	B585MLS	MCW Metrobus DR102/40	Alexander RL	H45/33F	1984	
887	B587MLS	MCW Metrobus DR132/3	Alexander RL	H45/33F	1984	Ex Kelvin Central, 1990
888	B588MLS	MCW Metrobus DR132/3	Alexander RL	H45/33F	1984	
893	B93PKS	MCW Metrobus DR102/47	Alexander RL	H45/33F	1984	Ex Kelvin Central, 1990
894	B94PKS	MCW Metrobus DR102/47	Alexander RL	H45/33F	1984	
895	B95PKS	MCW Metrobus DR102/47	Alexander RL	H45/33F	1984	
896	B96PKS	MCW Metrobus DR102/47	Alexander RL	H45/33F	1984	
898	B98PKS	MCW Metrobus DR132/6	Alexander RL	H45/33F	1985	
899	B99PKS	MCW Metrobus DR132/6	Alexander RL	H45/33F	1985	

Previous Registrations:

101ASV	BLS107Y	FSU380	ALS105Y	SSU827	MSC554X, WLT770, HGD711X
110ASV	D113ELS	FSU381	ALS101Y	SSU829	MSC553X, WLT760, HGD745X
119ASV	SSJ132Y	FSU382	ALS102Y	SSU831	B131PMS
143ASV	D112ELS	FSU383	ALS103Y	SSU837	B132PMS
144ASV	G170RBD	FSV634	D144HMS	SSU841	D141HMS
156ASV	D145HMS, 692FFC, D625GSG	GSU338	ERF72Y, 4327PL, FEH778Y	SSU857	D142HMS
365UMY	E617NLS	GSU339	ERF73Y, 8636PL, FEH780Y	SSU859	A129ESG, 692FFC
373GRT	E621NLS	HSU247	LPN355W	SSU861	MSC555X
692FFC	F914YNV	HSU273	LPN357W, 411DCD, OUF51W	SSU897	A168UGB
693AFU	A130ESG	HSU301	D114ELS	TSU682	F912YNV
7881UA	A118GLS	KSU834	D143HMS	TSV612	E620NLS
BSV807	A117GLS, WLT415, A253WYS	OVT798	A184UGB	VSU715	A124ESG
ESK958	F105SSE	PSU622	D51VSO	VXU444	E618NLS, FSU309, E771PSG
FSU302	RMS397W	PSU625	D54VSO	WLT724	E619NLS
FSU308	BLS108Y	PXI8935	USX969V	WSU479	A75JFA
FSU315	F412DUG	SSC108P	MSJ371P, 365UMY	WSU487	A21GBC
FSU318	RMS393W, FSU303	SSU816	A116GLS	WSU489	A22GBC
FSU334	BMS514Y	SSU821	A121ESG		

Livery: Blue and cream; yellow and blue (Scottish Citylink) 141-4, 204/8; red (Easyboarder) 663; white (National Express) 203/5; blue and gold (Bluebird Executive) 145, 202/7; red (Stirling District Council) 716.

MIDLAND RED WEST

Midland Red West Ltd, Heron Lodge, London Road, Worcester, WR5 2EW

Depots: Abbey Road, Evesham; Friar Street, Hereford; New Road, Kidderminster; Plymouth Road, Redditch and Padmore Street, Worcester. **Outstations** : Bull Ring bus station, Birmingham; Bishops Castle, Bridgnorth, Hopton Heath and Ludlow.

201-237
Dennis Lance 11SDA3107 Plaxton Verde B49F 1994

201	L201AAB	209	L209AAB	217	L217AAB	224	L224AAB	231	L231AAB
202	L202AAB	210	L210AAB	218	L218AAB	225	L225AAB	232	L232AAB
203	L203AAB	211	L211AAB	219	L219AAB	226	L226AAB	233	L233AAB
204	L204AAB	212	L212AAB	220	L220AAB	227	L227AAB	234	L234AAB
205	L205AAB	213	L213AAB	221	L221AAB	228	L228AAB	235	L235AAB
206	L206AAB	214	L214AAB	222	L322AAB	229	L229AAB	236	L236AAB
207	L207AAB	215	L215AAB	223	L223AAB	230	L230AAB	237	L237AAB
208	L208AAB	216	L216AAB						

238-256
Dennis Lance 11SDA3113 Plaxton Verde B49F 1995

238	M238MRW	242	M242MRW	246	M246MRW	250	M250MRW	254	M254MRW
239	M239MRW	243	M243MRW	247	M247MRW	251	M251MRW	255	M255MRW
240	M240MRW	244	M244MRW	248	M248MRW	252	M252MRW	256	M256MRW
241	M241MRW	245	M245MRW	249	M249MRW	253	M253MRW		

301-313
Dennis Dart 9.8SDL3054 Plaxton Pointer DP36F 1995

301	N301XAB	304	N304XAB	307	N307XAB	310	N310XAB	312	N312XAB
302	N302XAB	305	N305XAB	308	N308XAB	311	N311XAB	313	N313XAB
303	N303XAB	306	N306XAB	309	N309XAB					

539-658
Leyland National 11351A/1R   B49F 1976-77 Ex Midland Red, 1981

539	NOE539R	542	NOE542R	547	NOE547R	656	PUK656R	658	SOA658S
541	NOE541R	544	NOE544R	610	NOE610R	657	SOA657S		

Opposite, top: **A second batch of Dennis Lances with Plaxton Verde bodies arrived in 1995 and all nineteen are allocated to Worcester where they have replaced Leyland Nationals. Appropriate MRW registrations were obtained from Coventry VRO and 242, M242MRW is seen at Bromsgrove on the trunk 144 service from Birmingham.** *David Cole*
Opposite, bottom: **Another standard Badgerline order was for a further 13 Dennis Darts with Plaxton Pointer bodies received by Midland Red West in 1995 to replace elderly Leopards. They are fitted with high-back seating reflecting the use of some on the 69-mile long Hereford to Birmingham service 192 through Ludlow and Kidderminster, like 307, N307XAB seen here in Kidderminster.** *David Cole*

Since Midland Red West ceased operating regular National Express diagrams all vehicles used by the coach unit are now in red livery with Midland Red Coaches fleetnames. Seen on a private hire duty in Buckingham Palace Road is 1013, A678KDV one of three of the Plaxton Paramount 3500 in the fleet, this one originating with Devon General.
Colin Lloyd

Now down-graded for service work and wearing red and cream bus livery 853, Q553UOC is one of two Leyland Leopards rebodied by Plaxton in 1983-84 when only a few years old following accidents resulting in their Willowbrook bodies being beyond economic repair. This vehicle is seen turning into Gloucester bus station on service 373 from Worcester. *Malc McDonald*

722-752		Leyland National 11351A/1R				B49F	1978-79 Ex Midland Red, 1981		
722	WOC722T	**724**	WOC724T	**744**	XOV744T	**749**	XOV749T	**752**	XOV752T
723	WOC723T	**743**	XOV743T	**746**	XOV746T				

755	AFJ755T	Leyland National 11351A/1R	B50F	1979	Ex Western National, 1989
756	AFJ756T	Leyland National 11351A/1R	B50F	1979	Ex Western National, 1990
758	XOV758T	Leyland National 11351A/1R	B49F	1979	Ex Midland Red North, 1986
770	BVP770V	Leyland National 11351A/1R	B49F	1979	Ex Midland Red, 1981

774-783		Leyland Leopard PSU3E/4R		Plaxton Supreme IV		C49F*	1979-80 Ex Midland Red Coaches, 1986 *774/5/7 are C53F		
774	BVP774V	**776**	BVP776V	**778**	BVP778V	**782**	BVP782V	**783**	BVP783V
775	BVP775V	**777**	BVP777V	**781**	BVP781V				

853	Q553UOC	Leyland Leopard PSU3F/4R	Plaxton P'mount 3200 (1984) C49F	1982	Ex Midland Red Coaches, 1986
854	Q276UOC	Leyland Leopard PSU3E/4R	Plaxton P'mount 3200 (1983) C49F	1980	Ex Midland Red, 1981
1001	FEH1Y	Leyland Tiger TRCTL11/3R	Plaxton Paramount 3500 C50FT	1983	

1002-1007		Leyland Tiger TRCTL11/3RH		Plaxton Paramount 3200 II		C50FT*	1985	*1006/7 are C39FT	
1002	B102JAB	**1004**	B104JAB	**1005**	B105JAB	**1006**	B106JAB	**1007**	B107JAB
1003	B103JAB								

1008	LOA832X	Leyland Tiger TRCTL11/3R	Plaxton Supreme IV	C51F	1981	Ex Midland Red Coaches, 1986
1010	EAH890Y	Leyland Tiger TRCTL11/3R	Plaxton Paramount 3200 E	C53F	1983	Ex Midland Red Coaches, 1986
1011	A895KCL	Leyland Tiger TRCTL11/3R	Plaxton Paramount 3200 E	C53F	1983	Ex Midland Red Coaches, 1986
1012	A896KCL	Leyland Tiger TRCTL11/3R	Plaxton Paramount 3200 E	C53F	1983	Ex Midland Red Coaches, 1986
1013	A678KDV	Leyland Tiger TRCTL11/3R	Plaxton Paramount 3500	C48FT	1983	Ex Midland Red Coaches, 1986
1014	A656VDA	Leyland Tiger TRCTL11/3R	Plaxton Paramount 3500	C48FT	1983	Ex Midland Red Coaches, 1986
1015	A657VDA	Leyland Tiger TRCTL11/3R	Plaxton Paramount 3500	C48FT	1983	Ex Midland Red Coaches, 1986
1016	A658VDA	Leyland Tiger TRCTL11/3R	Plaxton Paramount 3200	C50FT	1983	Ex Midland Red Coaches, 1986

The Mercedes-Benz L608D was the standard minibus allocated by NBC to Midland Red West and the 104 received in 1985-86 permitted the conversion of most town services in Worcester, Kidderminster and Redditch. Initially each locality had its own livery but all are now in red and cream save for those carrying advertising liveries. Later second-hand purchases and some 609Ds saw the conversion of town services in Hereford, Malvern and Droitwich. A Robin Hood conversion 1321, C321PNP is seen in the latter town and it is envisaged that replacement of these vehicles will commence in 1996. *Colin Lloyd*

1017	B566BOK	Leyland Tiger TRCTL11/3RH	Duple Caribbean 2	C48FT	1984	Ex Midland Red Coaches, 1986
1018	B567BOK	Leyland Tiger TRCTL11/3RH	Duple Caribbean 2	C48FT	1984	Ex Midland Red Coaches, 1986
1019	B568BOK	Leyland Tiger TRCTL11/3RH	Duple Caribbean 2	C48FT	1984	Ex Midland Red Coaches, 1986
1020	C985HOX	Leyland Tiger TRCTL11/3RZ	Duple 340	C49FT	1986	Ex Midland Red Coaches, 1986
1021	C986HOX	Leyland Tiger TRCTL11/3RZ	Duple 340	C49FT	1986	Ex Midland Red Coaches, 1986
1022	C987HOX	Leyland Tiger TRCTL11/3RZ	Duple 340	C49FT	1986	Ex Midland Red Coaches, 1986

1101-1150   Leyland Lynx LX2R11C15Z4R   Leyland Lynx   B49F   1990

1101	G101HNP	1111	G111HNP	1121	G121HNP	1131	G131HNP	1141	G141HNP
1102	G102HNP	1112	G112HNP	1122	G122HNP	1132	G132HNP	1142	G142HNP
1103	G103HNP	1113	G113HNP	1123	G123HNP	1133	G133HNP	1143	G143HNP
1104	G104HNP	1114	G114HNP	1124	G124HNP	1134	G134HNP	1144	G144HNP
1105	G105HNP	1115	G115HNP	1125	G125HNP	1135	G135HNP	1145	G145HNP
1106	G106HNP	1116	G116HNP	1126	G126HNP	1136	G136HNP	1146	G146HNP
1107	G107HNP	1117	G117HNP	1127	G127HNP	1137	G137HNP	1147	G147HNP
1108	G108HNP	1118	G118HNP	1128	G128HNP	1138	G138HNP	1148	G148HNP
1109	G109HNP	1119	G119HNP	1129	G129HNP	1139	G139HNP	1149	G149HNP
1110	G110HNP	1120	G120HNP	1130	G130HNP	1140	G140HNP	1150	G150HNP

| 1300 | A670XUK | Mercedes-Benz L307D | Devon Conversions | M12 | 1984 | Ex Midland Red Coaches, 1986 |

1301-1319   Mercedes-Benz L608D   PMT Hanbridge   B20F*   1985   *1301-6 are DP20F

1301	C301PNP	1305	C305PNP	1309	C309PNP	1313	C313PNP	1317	C317PNP
1302	C302PNP	1306	C306PNP	1310	C310PNP	1314	C314PNP	1318	C318PNP
1303	C303PNP	1307	C307PNP	1311	C311PNP	1315	C315PNP	1319	C319PNP
1304	C304PNP	1308	C308PNP	1312	C312PNP	1316	C316PNP		

| 1320 | C320PNP | Mercedes-Benz L608D | Alexander | B20F | 1985 | |

1321-1361 — Mercedes-Benz L608D — Robin Hood — B20F — 1985-86

1321	C321PNP	1330	C330PNP	1338	C338PNP	1346	C346PNP	1354	C354PNP
1322	C322PNP	1331	C331PNP	1339	C339PNP	1347	C347PNP	1355	C355PNP
1323	C323PNP	1332	C332PNP	1340	C340PNP	1348	C348PNP	1356	C356PNP
1324	C324PNP	1333	C333PNP	1341	C341PNP	1349	C349PNP	1357	C357PNP
1325	C325PNP	1334	C334PNP	1342	C342PNP	1350	C350PNP	1358	C358PNP
1326	C326PNP	1335	C335PNP	1343	C343PNP	1351	C351PNP	1359	C359PNP
1327	C327PNP	1336	C336PNP	1344	C344PNP	1352	C352PNP	1360	C360PNP
1328	C328PNP	1337	C337PNP	1345	C345PNP	1353	C353PNP	1361	C361RUY
1329	C329PNP								

1362-1384 — Mercedes-Benz L608D — Reeve Burgess — B20F — 1986

1362	C362RUY	1367	C367RUY	1372	C372RUY	1377	C377RUY	1381	C381RUY
1363	C363RUY	1368	C368RUY	1373	C373RUY	1378	C378RUY	1382	C382RUY
1364	C364RUY	1369	C369RUY	1374	C374RUY	1379	C379RUY	1383	C383RUY
1365	C365RUY	1370	C370RUY	1375	C375RUY	1380	C380RUY	1384	C384RUY
1366	C366RUY	1371	C371RUY	1376	C376RUY				

1385-1404 — Mercedes-Benz L608D — Robin Hood — B20F — 1986

1385	C385RUY	1389	C389RUY	1393	C393RUY	1397	C397RUY	1401	C401RUY
1386	C386RUY	1390	C390RUY	1394	C394RUY	1398	C398RUY	1402	C402RUY
1387	C387RUY	1391	C391RUY	1395	C395RUY	1399	C399RUY	1403	C403RUY
1388	C388RUY	1392	C392RUY	1396	C396RUY	1400	C400RUY	1404	C404RUY

1405	A669XDA	Mercedes-Benz L508D	Reeve Burgess	DP17F	1984	Ex Midland Red Coaches, 1986

1406-1439 — Mercedes-Benz 609D — Reeve Burgess — B20F — 1987-88

1406	E406HAB	1413	E413KUY	1420	E420KUY	1427	E427KUY	1434	E434KUY
1407	E407HAB	1414	E414KUY	1421	E421KUY	1428	E428KUY	1435	E435KUY
1408	E408HAB	1415	E415KUY	1422	E422KUY	1429	E429KUY	1436	E436KUY
1409	E409HAB	1416	E416KUY	1423	E423KUY	1430	E430KUY	1437	E437KUY
1410	E410HAB	1417	E417KUY	1424	E424KUY	1431	E431KUY	1438	E438KUY
1411	E411HAB	1418	E418KUY	1425	E425KUY	1432	E432KUY	1439	E439KUY
1412	E412KUY	1419	E419KUY	1426	E426KUY	1433	E433KUY		

1440	C475BHY	Mercedes-Benz L608D	Reeve Burgess	B20F	1986	Ex Bristol Omnibus, 1988

1441-1449 — Mercedes-Benz L608D — Reeve Burgess — B20F — 1986 — Ex Southdown, 1988

1441	C581SHC	1443	C583SHC	1445	C585SHC	1447	C587SHC	1449	C589SHC
1442	C582SHC	1444	C584SHC	1446	C586SHC	1448	C588SHC		

1450	C788FRL	Mercedes-Benz L608D	Reeve Burgess	B20F	1986	Ex Western National, 1989
1451	C790FRL	Mercedes-Benz L608D	Reeve Burgess	B20F	1986	Ex Western National, 1989
1452	C207PCD	Mercedes-Benz L608D	Alexander	B20F	1986	Ex Brighton & Hove, 1990
1453	C208PCD	Mercedes-Benz L608D	Alexander	B20F	1986	Ex Brighton & Hove, 1990
1454	C209PCD	Mercedes-Benz L608D	Alexander	B20F	1986	Ex Brighton & Hove, 1990
1455	C212PCD	Mercedes-Benz L608D	Alexander	B20F	1986	Ex Brighton & Hove, 1990
1463	D763KWT	Mercedes-Benz 609D	Reeve Burgess	B20F	1987	Ex SWT, 1994

1476-1499 — Mercedes-Benz L608D — Reeve Burgess — B20F — 1986 — Ex City Line, 1993

1476	C476BHY	1487	C487BHY	1490	C490BHY	1492	C492BHY	1498	C498BHY
1477	C477BHY	1488	C488BHY	1491	C491BHY	1497	C497BHY	1499	C499BHY
1483	C483BHY								

1804	H204JHP	Peugeot-Talbot Pullman	Talbot	B22F	1990
1806	H206JHP	Peugeot-Talbot Pullman	Talbot	B22F	1990
1807	H207JHP	Peugeot-Talbot Pullman	Talbot	B22F	1990
1808	H208JHP	Peugeot-Talbot Pullman	Talbot	B22F	1990

Previous Registrations:

Q276UOC	BVP804V	Q553UOC	LOA843X

Livery: Red and cream; yellow and green (Quickstep) 1804-8; red (Midland Red Coaches) 1001-7/13-22, 1300; yellow, red and black (Midland Express) 775-8/81-3

Fifty Leyland Lynx were delivered in 1990 and were the first new, full-size, single decks for ten years. They were initially allocated to Digbeth for Birmingham area services, then replaced by the first batch of Dennis Lance and now reallocated to Redditch and Kidderminster. Seen leaving Birmingham bus station for Blackheath is 1135, G135HNP. *Malc McDonald*

The 1995 intake of Dennis Lance has seen an inroad into the remaining Leyland Nationals and most of those still in use are based at Evesham and Hereford. In both 1994 and 1995 West Midlands Travel bought most of the redundant examples. A former Worcester garage vehicle 722, WOC722T is now at Hereford but seen here about to cross the River Severn at Worcester. *Ken Jubb*

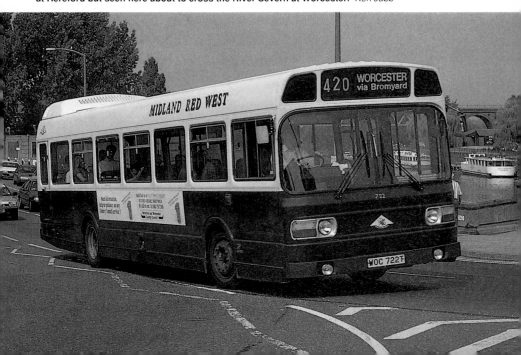

The latest model to enter FirstBus operation is the recently introduced Volvo B10L model that represents a new standard for Citybuses and includes a low floor and low emissions. Bodywork on the model, as shown as Northampton 41, N41RRP, is the Alexander Ultra, a new product based on the Säffle design and using System 2000 body structure which is exclusive to Alexander and features the proven heavy duty bolted aluminium construction system ideal for the demands of the low floor layout here featuring a no-step gangway over 80% of the vehicle length. In contrast is an earlier version of the Volvo Citybus, based on a B10M chassis and also with an Alexander body, in this case the double-deck RV-type. Photographed in Northampton is 90, H290VRP which has high-back seating in the lower saloon.

NORTHAMPTON TRANSPORT

Northampton Transport Ltd, The Bus Depot, St James Road, Northampton, NN5 5JD

41	N41RRP	Volvo B10L		Alexander Ultra		B45F	1995	
42	N42RRP	Volvo B10L		Alexander Ultra		B45F	1995	
43	N43RRP	Volvo B10L		Alexander Ultra		B45F	1995	

48-58		Bristol VRT/SL3/6LXB		Alexander AL		H45/27D	1978	
48	CNH48T	50	CNH50T	53	CNH53T	55	CNH55T	58	CNH58T
49	CNH49T	52	CNH52T	54	CNH54T	57	CNH57T		

70	VVV70S	Bristol VRT/SL3/6LXB		Alexander AL		H45/28D	1977

74	ABD74X	Bristol VRT/SL3/6LXB	East Lancashire	H44/28D	1982
75	ABD75X	Bristol VRT/SL3/6LXB	East Lancashire	DPH43/27D	1982
76	ABD76X	Bristol VRT/SL3/6LXB	East Lancashire	DPH43/27D	1982

83-88		Volvo Citybus B10M-50		Alexander RV		DPH47/35F	1989	
83	F83XBD	85	F85XBD	86	F86DVV	87	F87DVV	88	F88DVV
84	F84XBD							

89-94		Volvo Citybus B10M-55		Alexander RV		DPH47/35F	1990	
89	H289VRP	91	H291VRP	92	H292VRP	93	H293VRP	94	H294VRP
90	H290VRP							

95-100		Volvo Citybus B10M-50		Alexander RV		DPH47/35F	1991	
95	J295GNV	97	J297GNV	98	J298GNV	99	J299GNV	100	J210GNV
96	J296GNV							

101	D101XNV	Volvo Citybus B10M-50	East Lancashire	DPH45/31F	1986
102	D102XNV	Volvo Citybus B10M-50	East Lancashire	DPH45/31F	1986
111	E111NNV	Volvo Citybus B10M-50	Duple 300	DP49F	1988
112	G112ENV	Volvo Citybus B10M-55	Duple 300	DP49F	1989
113	G113ENV	Volvo Citybus B10M-55	Duple 300	DP51F	1989
114	G114ENV	Volvo Citybus B10M-55	Duple 300	DP51F	1989
115	J115MRP	Volvo Citybus B10M-55	East Lancashire	DP47F	1992

121-132		Volvo Citybus B10M-50		Alexander RV		H47/35F	1992-93	
121	K121URP	124	K124URP	127	K127GNH	129	K129GNH	131	K131GNH
122	K122URP	125	K125URP	128	K128GNH	130	K130GNH	132	K132GNH
123	K123URP	126	K126URP						

Livery: Cream, red and maroon

Named vehicles:
83 *Northampton Charter 1189*, 84 *Northampton Castle*, 85 *Richard the Lionheart*, 86 *Queen Eleanor*, 87 *Danes Camp*, 88 *Simon De Senlis*, 89 *Delapre Abbey*, 90 *Rush Mills - Penny Black*, 91 *Knights Templar*, 92 *Master Cobbler*, 93 *Nene Navigation*, 94 *Philip Doddridge*, 95 *Thomas Chipsey - 1544*, 96 *Becketts Well*, 97 *William Carey 1761-1834*, 98 *John Clare 1793-1864*, 99 *Charles Bradlaugh 1833-1891*, 100 *All Saints Church*, 121 *Sir Phillip Manfield*, 122 *Hazelrigg House*, 123 *The Guildhall*, 124 *The Welsh House*, 125 *Notre Dame*, 126 *Lt Col Mobbs DSO*, 127 *A C Parkhouse*, 128 *Grand Junction Canal*, 129 *Market Square*, 130 *Sixfields*, 131 *Rose of the Shires*, 132 *Mary, Queen of Scots*.

PEOPLE'S PROVINCIAL

The Provincial Bus Company Ltd, Hoeford, Gosport Road, Fareham PO16 0ST

1	A301KJT	Leyland National 2 NL116L11/1R				DP47F	1984	
2	A302KJT	Leyland National 2 NL116L11/1R				DP47F	1984	
3	H523CTR	Ace Cougar		Wadham Stringer Portsdown		B41F	1990	

13-22
Leyland National 1151/2R/0403 — B44D — 1972-74

13	HOR413L	15	HOR415L	17u	HOR417L	19	PCG919M	21	PCG921M
14	HOR414L	16	HOR416L	18	PCG918M	20	PCG920M	22	PCG922M

23	RUF37R	Leyland National 11351A/2R				B44D	1977	Ex Rennies, Dunfirmline, 1988

26-33
Leyland National 11351/2R — B44D — 1974-75

26u	UAA226M	28	GCR728N	30u	JBP130P	32	JBP132P	33	JBP133P
27	GCR727N	29	JBP129P	31	JBP131P				

34-44
Leyland National 11351A/2R — B44D — 1976-79

34	LTP634R	36	MOW636R	38	PTR238S	40	SBK740S	42	TPX42T
35u	LTP635R	37	MOW637R	39u	PTP239S	41	TPX41T	44	UPO444T

71	UFX848S	Leyland National 11351A/1R				B49F	1977	Ex Hants & Dorset, 1983
72	VFX980S	Leyland National 11351A/1R				B49F	1978	Ex Hants & Dorset, 1983
73	EEL893V	Leyland National 11351A/1R(Volvo)				DP52F	1979	Ex Hants & Dorset, 1983
100	BDL65T	Bedford YMT		Plaxton Supreme IV		C46F	1979	Ex Solent Blue Line, 1993

118-127
Iveco Daily 49.10 — Robin Hood City Nippy — B19F* — 1986 — *127 is B21F

118	D118DRV	120	D120DRV	122	D122DRV	123	D123DRV	127	D127DRV
119	D119DRV	121u	D121DRV						

128-136
Iveco Daily 49.10 — Robin Hood City Nippy — B24F — 1989

128	F128SBP	130	F130SBP	132	F132SBP	134	F134TCR	136	F136TCR
129	F129SBP	131	F131SBP	133	F133TCR	135	F135TCR		

The rare Ace Cougar with Wadham Stringer Portsdown bodywork is now number 3 in the People's Provincial fleet. This interesting vehicle was displayed at Showbus in September 1995 where this picture was taken.
Mark Willis

No.73 in the People's Provincial fleet is the only remaining Mark 1 National to be supplied to the Suburban Express specification. This bus has been fitted with a Volvo engine which conforms to Euro2 emission requirements. Registered EEL893V, it was photographed while working service 9 to HMS Dolphin at Gosport. *Andrew Jarosz*

Representing the minibus fleet of People's Provincial is 142, J142KPX. An Iveco 49.10 with Carlyle Dailybus 2 bodywork, it carries a livery of two-tone green and white roof. As the book was being prepared, the final date for FirstBus to take control of this fleet was still not settled. *Andrew Jarosz*

Among the first pictures of the new Urban Star body is this of People's Provincial 601, taken at the coachworks, then unregistered. It made its debut at the 1995 Bus and Coach show in October when the manufacturers, UVG Buses announced their intention to drop the WS Coachbuilders name although this will not happen overnight. *Mike Willis*

137	G137WOW	Iveco Daily 49.10	Phoenix	B24F	1989	
138	G138WOW	Iveco Daily 49.10	Phoenix	B24F	1989	
139	G139WOW	Iveco Daily 49.10	Phoenix	B24F	1989	

140-146   Iveco Daily 49.10   Carlyle   B23F   1992

140	J140KPX	142	J142KPX	144	J144KPX	145	J145KPX	146	J146KPX
141	J141KPX	143	J143KPX						

160-165   Iveco TurboDaily 59.12   WS Coachbuilders Daily Bus B27F   1993

160	K160PPO	162	K162PPO	163	K163PPO	164	K164PPO	165	K165PPO
161	K161PPO								

166-207   Iveco TurboDaily 59.12   Marshall C31   B27F   1994-95

166	L166TRV	175	L175TRV	183	M183XTR	191	M191XTR	199	M199XTR
167	L167TRV	176	L176TRV	184	M184XTR	192	M192XTR	201	M201XTR
168	L168TRV	177	L177TRV	185	M185XTR	193	M193XTR	202	M202XTR
169	L169TRV	178	L178TRV	186	M186XTR	194	M194XTR	203	M203XTR
170	L170TRV	179	M179XTR	187	M187XTR	195	M195XTR	204	M204BPO
171	L171TRV	180	M180XTR	188	M188XTR	196	M196XTR	205	M205BPO
172	L172TRV	181	M181XTR	189	M189XTR	197	M197XTR	206	M206BPO
173	L173TRV	182	M182XTR	190	M190XTR	198	M198XTR	207	M207BPO
174	L174TRV								

303	NFN79M	Leyland National 1151/1R/2402	B49F	1974	Ex National Welsh, 1987
305	NPD146L	Leyland National 1151/1R/0402	B49F	1973	Ex London Country, 1983
306	NPD154L	Leyland National 1151/1R/0402	B49F	1973	Ex London Country, 1983
343	UPO443T	Leyland National 11351A/1R	B44D	1979	
363	MJT880P	Leyland National 11351/1R	B49F	1976	
365	RJT147R	Leyland National 11351A/1R	B49F	1977	Ex Hants & Dorset, 1983
366	RJT148R	Leyland National 11351A/1R	B49F	1977	Ex Hants & Dorset, 1983
367	SPR39R	Leyland National 11351A/1R	B49F	1977	Ex Hants & Dorset, 1983
368	SPR40R	Leyland National 11351A/1R	B49F	1977	Ex Hants & Dorset, 1983
369	SPR41R	Leyland National 11351A/1R	B49F	1977	Ex Hants & Dorset, 1983

The two Leyland Fleetlines in the People's Provincial fleet were augmented during the last few years with several VRTs from other FirstBus fleets. Illustrating the Fleetlines is 592, NFX131P, seen here with its roof fitted - the vehicle now forming part of the open top trio. *Andrew Jarosz*

370	UFX847S	Leyland National 11351A/1R			B49F	1977	Ex Hants & Dorset, 1983	
375	WFX253S	Leyland National 11351A/1R			DP48F	1978	Ex Hants & Dorset, 1983	
376	WFX257S	Leyland National 11351A/1R			DP48F	1978	Ex Hants & Dorset, 1983	
383	NOE561R	Leyland National 11351A/1R			B49F	1976	Ex Midland Red East, 1983	
394	BCD824L	Leyland National 1151/1R/0102			B49F	1973	Ex Southdown, 1988	

401-414

		Leyland National 10351A/2R			B36D	1976-79 Ex London Buses, 1991

401	KJD528P	404	AYR331T	407	THX234S	410	THX115S	413	AYR341T
402	THX248S	405	AYR344T	408	BYW415V	411	THX131S	414	YYE276T
403	AYR299T	406	YYE278T	409	KJD511P	412	THX242S		

507	LWU471V	Bristol VRT/SL3/6LXB	Eastern Coach Works	H39/31F	1980	Ex Rider York, 1995
508	UVX2S	Bristol VRT/SL3/6LXB	Eastern Coach Works	H39/31F	1977	Ex Badgerline, 1994
509	SFJ101R	Bristol VRT/SL3/6LXB	Eastern Coach Works	H43/31F	1977	Ex Western National, 1993
510	UTO836S	Bristol VRT/SL3/6LXB	Eastern Coach Works	H43/31F	1977	Ex Western National, 1993
511	AFJ748T	Bristol VRT/SL3/6LXB	Eastern Coach Works	H43/31F	1979	Ex Western National, 1993
512	AFJ752T	Bristol VRT/SL3/6LXB	Eastern Coach Works	H43/31F	1979	Ex Western National, 1993
513	AFJ763T	Bristol VRT/SL3/6LXB	Eastern Coach Works	H43/31F	1979	Ex Western National, 1993
515	NTC573R	Bristol VRT/SL3/6LXB	Eastern Coach Works	H43/27D	1977	Ex City Line, 1994
516	RHT503S	Bristol VRT/SL3/6LXB	Eastern Coach Works	H43/27D	1978	Ex City Line, 1994
517	RHT504S	Bristol VRT/SL3/6LXB	Eastern Coach Works	H43/27D	1978	Ex City Line, 1994
518	RHT512S	Bristol VRT/SL3/6LXB	Eastern Coach Works	H43/27D	1978	Ex City Line, 1994
519	TWS908T	Bristol VRT/SL3/6LXB	Eastern Coach Works	H43/27D	1979	Ex City Line, 1994
520	AHU514V	Bristol VRT/SL3/6LXB	Eastern Coach Works	H43/27D	1980	Ex City Line, 1994
591	NFX130P	Leyland Fleetline FE30LR	Alexander AD	CO43/31F	1976	Ex Yellow Buses, 1991
592	NFX131P	Leyland Fleetline FE30LR	Alexander AD	CO43/31F	1976	Ex Yellow Buses, 1991
593	MOD571P	Bristol VRT/SL3/6LXB	Eastern Coach Works	O43/31F	1976	Ex Western National, 1993

601-607

		Dennis Dart 9.8SDL3054	UVG Urban Star	B40F	1995

601	N601EBP	603	N603EBP	605	N605EBP	606	N606EBP	607	N607EBP
602	N602EBP	604	N604EBP						

Livery: Cream and dark green

PMT

PMT Ltd, 33 Woodhouse Street, Stoke-on-Trent ST4 1EQ

Depots: Scotia Road, Burslem; Brookhouse Industrial Estate, Cheadle; Liverpool Road, Chester; Second Avenue, Crewe Gates Farm, Crewe; Platt Street, Dukinfield; Bus Station, Ellesmere Port; Pasture Road, Moreton; Liverpool Road, Newcastle-under-Lyme and New Chester Road, Rock Ferry.

MXU22	H202JHP	Peugeot-Talbot Pullman	Talbot	B22F	1990	Ex Midland Red West, 1995
MXU23	H203JHP	Peugeot-Talbot Pullman	Talbot	B22F	1990	Ex Midland Red West, 1995
STL24	ERF24Y	Leyland Tiger TRCTL11/3R	Plaxton Paramount 3500	C53F	1983	
MBU25	M25YRE	Peugeot Boxer	TBP	M9	1995	
MBU26	M26YRE	Peugeot Boxer	TBP	M9	1995	
MBU27	M27YRE	Peugeot Boxer	TBP	M9	1995	
MBU28	M28YRE	Peugeot Boxer	TBP	M9	1995	
STL43	E43JRF	Leyland Tiger TRCTL11/3R	Plaxton Paramount 3500 III	C53F	1988	
STL44	E44JRF	Leyland Tiger TRCTL11/3R	Plaxton Paramount 3500 III	C53F	1988	
MMM50	B232AFV	Mercedes-Benz L307D	Reeve Burgess	M12	1985	Ex Landliner, Birkenhead, 1990
SLL80	NED433W	Leyland Leopard PSU3E/4R	Plaxton Supreme IV	C53F	1981	Ex Turner, Brown Edge, 1988
MMM88	C108SFP	Mercedes-Benz L307D	Reeve Burgess	M12	1985	Ex Goldcrest, Birkenhead, 1990
MMM89	F660EBU	Mercedes-Benz 609D	North West Coach Sales	B19F	1988	Ex Landliner, Birkenhead, 1990
MMM97	D176VRP	Mercedes-Benz L608D	Alexander	B20F	1986	Ex Milton Keynes Citybus, 1992
MMM98	D185VRP	Mercedes-Benz L608D	Alexander	B20F	1986	Ex Milton Keynes Citybus, 1992
MMM99	C683LGE	Mercedes-Benz L608D	Reeve Burgess	B20F	1985	Ex Strathclyde, 1991
MMM100	F100UEH	Mercedes-Benz 609D	PMT	C24F	1989	
MMM101	G101EVT	Mercedes-Benz 609D	PMT	C21F	1990	
MMM102	F452YHF	Mercedes-Benz 811D	North West Coach Sales	C24F	1989	Ex C & M, Aintree, 1992
MMM104	F713OFH	Mercedes-Benz 307D	North West Coach Sales	M12	1989	Ex van, 1992
IFF105	J328RVT	Iveco Daily 49.10	Reeve Burgess Beaver	C29F	1991	Ex Roseville Taxis, Newcastle, 1993
MRP106	E106LVT	Renault-Dodge S56	PMT	C22F	1988	
MMM107	YRF267X	Mercedes-Benz L307D	Devon Conversions	M12	1982	Ex Kelly, Lower Gornal, 1985
MRP108	D162LTA	Renault-Dodge S56	Reeve Burgess	B23F	1987	Ex Cardiff Bus, 1994
MMM109	F217OFB	Mercedes-Benz 307D	North West Coach Sales	M12L	1989	Ex van, 1992
IMM110	H189CNS	Mercedes-Benz 814D	Dormobile Routemaker	C33F	1991	Ex Executive Travel, 1994
MMM112	XRF2X	Mercedes-Benz L307D	Reeve Burgess	M12	1983	
MMM114	G805AAD	Mercedes-Benz 308	North West Coach Sales	M12L	1989	Ex van, 1992
MMM115	XRF1X	Mercedes-Benz L608D	PMT Hanbridge	C21FL	1984	
MMM116	FXI8653	Mercedes-Benz L608D	PMT Hanbridge	C21FL	1984	
MMM117	B117OBF	Mercedes-Benz L608D	PMT Hanbridge	B19F	1984	
MMM118	C684LGE	Mercedes-Benz L608D	Alexander AM	B20F	1986	Ex Strathclyde, 1991
MMM119	B119RRE	Mercedes-Benz L608D	PMT Hanbridge	B19F	1985	

PMT introduced the Crosville-style prefix to the fleet number system though there is no duplication of numbers in the series. A full breakdown of these is given in the North Midlands Bus Handbook and represent type, chassis and engine. Shown here is STL43, E43JRF, a service bus Tiger with Leyland engine.
Cliff Beeton

MMM120-159 — Mercedes-Benz L608D — PMT Hanbridge — B20F — 1985-86

120	C120VBF	128	C128VRE	137	C137VRE	145	C145WRE	153	D153BEH
121	C121VRE	130	C130VRE	138	C138VRE	146	C146WRE	154	D154BEH
122	C122VRE	131	C131VRE	139	C139VRE	147	C147WRE	155	D155BEH
123	C123VRE	132	C132VRE	140	C140VRE	148	C148WRE	156	D156BEH
124	C124VRE	133	C133VRE	141	C141VRE	149	C149WRE	157	D157BEH
125	C125VRE	134	C134VRE	142	C142VRE	150	C150WRE	158	D158BEH
126	C126VRE	135	C135VRE	143	C143VRE	151	C151WRE	159	D159BEH
127	C127VRE	136	C136VRE	144	C144VRE	152	D152BEH		

MFF178	F166ONT	Ford Transit VE6	Dormobile	M15	1989	Ex Shropshire CC, 1994	

MMM180-189 — Mercedes-Benz L608D — PMT Hanbridge — B20F — 1986

180	D180BEH	183	D183BEH	185	D185BEH	187	D187BEH	189	D189BEH
182	D182BEH	184	D184BEH	186	D186BEH	188	D188BEH		

MMM190	C124LHS	Mercedes-Benz L608D	Reeve Burgess	B20F	1986	Ex Strathclyde, 1991
MMM191	D118PGA	Mercedes-Benz L608D	Reeve Burgess	B19F	1986	Ex Strathclyde, 1991
MMM192	D119PGA	Mercedes-Benz L608D	Imperial	B23F	1986	Ex Strathclyde, 1991

MMM193-197 — Mercedes-Benz L608D — PMT Hanbridge — B19F — 1986 — Ex Strathclyde, 1991

193	D120PGA	194	D121PGA	195	D122PGA	196	D123PGA	197	D124PGA

MMM199-209 — Mercedes-Benz L608D — Alexander — B20F* — 1986 — Ex Milton Keynes Citybus, 1992

*204/5 are B19F

199	D159VRP	202	D165VRP	204	D180VRP	206	D187VRP	208	D178VRP
200	D160VRP	203	D179VRP	205	D186VRP	207	D157VRP	209	D184VRP
201	D162VRP								

MMM210	C706JMB	Mercedes-Benz L609D	Reeve Burgess	B19F	1986	Ex Crosville, 1990
MMM211	C709JMB	Mercedes-Benz L609D	Reeve Burgess	B19F	1986	Ex Crosville, 1990
MMM212	C710JMB	Mercedes-Benz L609D	Reeve Burgess	B19F	1986	Ex Crosville, 1990
MMM213	C711JMB	Mercedes-Benz L609D	Reeve Burgess	B19F	1986	Ex Crosville, 1990
MMM215	D548FAE	Mercedes-Benz L608D	Dormobile	B20F	1986	Ex City Line, 1994
MMM216	D549FAE	Mercedes-Benz L608D	Dormobile	B20F	1986	Ex City Line, 1994
MMM217	D550FAE	Mercedes-Benz L608D	Dormobile	B20F	1986	Ex City Line, 1994
MMM218	D551FAE	Mercedes-Benz L608D	Dormobile	B20F	1986	Ex City Line, 1994

MPC224-230 — MCW MetroRider MF150/118 — MCW — B25F — 1988 — Ex Crosville, 1990

224	F88CWG	226	F106CWG	228	F108CWG	229	F109CWG	230	F110CWG
225	F95CWG	227	F107CWG						

MPC231	L231NRE	Optare MetroRider	Optare	B31F	1994

1994 saw the addition of seven Optare MetroRiders to the PMT fleet. Photographed heading for Trentham is IPC383. The I code was introduced to designate the midi-bus in between the mini and the full size single deck service bus.
Cliff Beeton

SNL287	SFA287R	Leyland National 11351A/1R				B52F	1977	
SNG297	EMB358S	Leyland National 11351A/1R				B49F	1978	Ex Crosville, 1990
SNG298	GMB377T	Leyland National 11351A/1R				B49F	1978	Ex Crosville, 1990

IWC310-318

		Leyland Swift LBM6T/2RS		PMT Knype		DP37F*	1988-89 *315-8 are DP35F	

310	F310REH	312	F312REH	315	F315REH	317	F317REH	318	G318YVT
311	F311REH	313	F313REH	316	F316REH			

IWC320	E342NFA	Leyland Swift LBM6T/2RS	PMT Knype	DP37F	1988	Ex PMT demonstrator, 1988
IPC321	L321HRE	Optare MetroRider	Optare	DP30F	1993	
IPC322	L269GBU	Optare MetroRider	Optare	B29F	1993	
IPC323	L323NRF	Optare MetroRider	Optare	B29F	1994	

IMM330-353

		Mercedes-Benz 811D		PMT Ami		B28F	1989-90

330	G330XRE	335	G335XRE	340	G340XRE	345	G345CBF	350	G550ERF
331	G331XRE	336	G336XRE	341	G341XRE	346	G346CBF	351	H351HRF
332	G332XRE	337	G337XRE	342	G342CBF	347	G347ERF	352	H352HRF
333	G333XRE	338	G338XRE	343	G343CBF	348	G348ERF	353	H353HRF
334	G334XRE	339	G339XRE	344	G344CBF	349	G349ERF		

IMM354	H354HVT	Mercedes-Benz 811D	Reeve Burgess Beaver	B31F	1990	
IMM355	H355HVT	Mercedes-Benz 811D	Reeve Burgess Beaver	B31F	1990	
IMM356	H356HVT	Mercedes-Benz 811D	Reeve Burgess Beaver	B33F	1990	
IMM357	H357HVT	Mercedes-Benz 811D	Reeve Burgess Beaver	B33F	1990	

IMM358-363

		Mercedes-Benz 811D	PMT Ami	B29F	1990

358	H358JRE	360	H160JRE	361	H361JRE	362	H362JRE	363	H363JRE
359	H359JRE								

MMM364	E39KRE	Mercedes-Benz L811D	PMT	B25F	1988	Ex van, 1990
IMM365	G495FFA	Mercedes-Benz 811D	PMT Ami	B28F	1990	

IMM366-371

		Mercedes-Benz 811D	PMT Ami	B29F	1991

366	H366LFA	368	H368LFA	369	H369LFA	370	H370LFA	371	H371LFA
367	H367LFA								

IMM372	H372MEH	Mercedes-Benz 811D	Whittaker-Europa	B31F	1991	
IMM373	H373MVT	Mercedes-Benz 811D	PMT Ami	B29F	1991	
IMM374	K374BRE	Mercedes-Benz 811D	Autobus Classique	B29F	1992	
IMM375	K375BRE	Mercedes-Benz 811D	Autobus Classique	B29F	1992	
IMM376	J920HGD	Mercedes-Benz 709D	Dormobile Routemaker	B29F	1991	Ex Stonier, 1994

IPC377-383

		Optare MetroRider	Optare	B29F	1994

377	M377SRE	379	M379SRE	381	M381SRE	382	M382SRE	383	M383SRE
378	M378SRE	380	M380SRE						

MMM430-448

		Mercedes-Benz 709D	Plaxton Beaver	B24F	1992

430	J430WFA	434	K434XRF	438	K438XRF	442	K442XRF	446	K446XRF
431	J431WFA	435	K435XRF	439	K439XRF	443	K443XRF	447	K447XRF
432	K432XRF	436	K436XRF	440	K440XRF	444	K544XRF	448	K448XRF
433	K433XRF	437	K437XRF	441	K441XRF	445	K445XRF		

MXU449	K449XRF	Peugeot-Talbot Pullman	TBP	B18F	1992	
MMM450	E791CCA	Mercedes-Benz L507D	PMT	B20F	1988	Ex Roberts, Bootle, 1992

Opposite, top: **The first Dennis Lance for the PMT fleet was SDC862, L862HFA, an example with Northern Counties Paladin bodywork which joined the fleet in 1993. It carries Crosville names and is based on the Wirral where it was photographed. A further five Lance were being delivered as the book was being prepared.** *Cliff Beeton*
Opposite, bottom: **Dennis Darts with both Plaxton and Marshall bodywork are used by PMT. The latest examples carry the Pointer design as illustrated by IDC957, M957XVT, as it heads for Trentham.** *Cliff Beeton*

MMM451-466   Mercedes-Benz L608D   PMT Hanbridge   B20F   1987-88

451	D451ERE	454	D454ERE	457	D457ERE	460	E760HBF	464	E764HBF
452	D452ERE	455	D455ERE	458	D458ERE	461	E761HBF	465	E765HBF
453	D453ERE	456	D456ERE	459	D459ERE	462	E762HBF	466	E766HBF

MMM467	E767HBF	Mercedes-Benz 709D	PMT	B21F	1988
MMM468	E768HBF	Mercedes-Benz 609D	PMT	B20F	1987
MMM469	E769HBF	Mercedes-Benz 609D	Reeve Burgess	B20F	1987
MMM470	E470MVT	Mercedes-Benz 709D	PMT	B23F	1988
MMM471	E471MVT	Mercedes-Benz 609D	PMT	B20F	1988
MMM472	F472RBF	Mercedes-Benz 609D	PMT	B20F	1988
MMM473	F473RBF	Mercedes-Benz 609D	PMT	B20F	1988
MMM474	E41JRF	Mercedes-Benz 709D	PMT	B23F	1988
MMM475	F475VEH	Mercedes-Benz 609D	PMT	B20F	1989
MMM476	E831ETY	Mercedes-Benz 609D	Reeve Burgess Beaver	B20F	1988	Ex Vasey, Ashington, 1990
MMM477	G477ERF	Mercedes-Benz 609D	PMT	B20F	1990
MMM478	G478ERF	Mercedes-Benz 609D	PMT	B20F	1990
MMM479	E384XCA	Mercedes-Benz 609D	PMT	B24F	1987	Ex Dennis's, Ashton, 1990
MMM480	H180JRE	Mercedes-Benz 709D	PMT	B20F	1990
MMM481	H481JRE	Mercedes-Benz 709D	PMT	B25F	1990
MMM482	H482JRE	Mercedes-Benz 609D	Whittaker Europa	B20F	1990
MMM483	H483JRE	Mercedes-Benz 609D	Whittaker Europa	B20F	1990
MMM484	J484PVT	Mercedes-Benz 709D	PMT	B25F	1991
MMM485	J485PVT	Mercedes-Benz 709D	Whittaker (PMT)	B25F	1992
MMM486	J486PVT	Mercedes-Benz 709D	Whittaker (PMT)	B25F	1992

MMM487-498   Mercedes-Benz 709D   Dormobile Routemaker   B24F*   1993   *488/9 are B27F

487	K487CVT	490	K490CVT	493	L493HRE	495	L495HRE	497	L497HRE
488	K488CVT	491	K491CVT	494	L494HRE	496	L496HRE	498	L498HRE
489	K489CVT	492	K492CVT						

MRP501	E801HBF	Renault-Dodge S56	PMT	B25F	1987

MRP502-528   Renault-Dodge S56   Alexander AM   B20F*   1987   *527/8 are B25F

502	E802HBF	508	E808HBF	513	E813HBF	518	E818HBF	524	E824HBF
503	E803HBF	509	E809HBF	514	E814HBF	519	E819HBF	525	E825HBF
504	E804HBF	510	E810HBF	515	E815HBF	520	E820HBF	526	E826HBF
505	E805HBF	511	E811HBF	516	E816HBF	521	E821HBF	527	E527JRE
506	E806HBF	512	E812HBF	517	E817HBF	522	E822HBF	528	E528JRE
507	E807HBF								

MRP530	E526NEH	Renault-Dodge S56	PMT	B25F	1988	
MRP531	F531UVT	Renault-Dodge S56	PMT	B25F	1989	
MRP532	G532CVT	Renault-Dodge S56	PMT	B25F	1990	
MMM550	H834GLD	Mercedes-Benz 609D	North West Coach Sales	B19F	1990	Ex Capital, West Drayton, 1994
MMM551	H835GLD	Mercedes-Benz 609D	North West Coach Sales	B19F	1990	Ex Capital, West Drayton, 1994
MMM552	H836GLD	Mercedes-Benz 609D	North West Coach Sales	B19F	1990	Ex Capital, West Drayton, 1994

MMM553-563   Mercedes-Benz 709D   Marshall C19   B24F   1994

553	L553LVT	556	L556LVT	558	L558LVT	560	M660SRE	562	M562SRE
554	L554LVT	557	L557LVT	559	M559SRE	561	M561SRE	563	M563SRE
555	L455LVT								

MMM564-573   Mercedes-Benz 709D   Plaxton Beaver   B24F   1994

564	M564SRE	566	M566SRE	568	M568SRE	570	M570SRE	572	M572SRE
565	M565SRE	567	M567SRE	569	M569SRE	571	M571SRE	573	M573SRE

MMM574-594   Mercedes-Benz 709D   Plaxton Beaver   B22F   1995

574	N574CEH	579	N579CEH	583	N583CEH	587	N587CEH	591	N591CEH
575	N575CEH	580	N580CEH	584	N584CEH	588	N588CEH	592	N592CEH
576	N576CEH	581	N581CEH	585	N585CEH	588	N589CEH	593	N593CEH
577	N577CEH	582	N582CEH	586	N586CEH	590	N590CEH	594	N594CEH
578	N578CEH								

Representing the Bristol VRT in the PMT fleet, of which some fifty remain, is DVL718, MFA718V. The PMT fleet was one of the few to receive VRs to the low height of 13'5" while the most common height was 13'8". PMT also chose Leyland engines for their vehicles. *Cliff Beeton*

DVG606	WDM341R	Bristol VRT/SL3/501(6LXB)	Eastern Coach Works	H43/31F	1977	Ex Crosville, 1990
DVG607	UDM450V	Bristol VRT/SL3/501(6LXB)	Eastern Coach Works	H43/31F	1980	Ex Crosville, 1990
DVG608	VCA452W	Bristol VRT/SL3/501(6LXB)	Eastern Coach Works	H43/31F	1980	Ex Crosville, 1990
DVG609	VCA464W	Bristol VRT/SL3/501(6LXB)	Eastern Coach Works	H43/31F	1980	Ex Crosville,.1990
DVG610	WTU465W	Bristol VRT/SL3/501(6LXB)	Eastern Coach Works	H43/31F	1980	Ex Crosville, 1990
DVG611	WTU472W	Bristol VRT/SL3/501(6LXB)	Eastern Coach Works	H43/31F	1980	Ex Crosville, 1990
DVG612	WTU481W	Bristol VRT/SL3/501(6LXB)	Eastern Coach Works	H43/31F	1981	Ex Crosville, 1990
DVG613	WTU482W	Bristol VRT/SL3/501(6LXB)	Eastern Coach Works	H43/31F	1981	Ex Crosville, 1990
DVG614	WTU483W	Bristol VRT/SL3/501(6LXB)	Eastern Coach Works	H43/31F	1981	Ex Crosville, 1990
DVG615	DCA526X	Bristol VRT/SL3/501(6LXB)	Eastern Coach Works	H43/31F	1981	Ex Crosville, 1990

DVG616-621

	Bristol VRT/SL3/6LXB	Eastern Coach Works	H43/31F	1979	Ex Thames Transit, 1989

616	YBW487V	**618**	YBW489V	**619**	EJO490V	**620**	EJO491V	**621**	EJO492V

DVG622	507EXA	Bristol VRT/SL2/6G	Eastern Coach Works	O43/31F	1974	
DVG624	AHU515V	Bristol VRT/SL3/6LXB	Eastern Coach Works	H43/27D	1980	Ex City Line, 1994
DVG625	AHW203V	Bristol VRT/SL3/6LXB	Eastern Coach Works	H43/27D	1980	Ex City Line, 1994
DVL635	UMB333R	Bristol VRT/SL3/501	Eastern Coach Works	H43/31F	1977	Ex Crosville, 1990
DVL640	ODM409V	Bristol VRT/SL3/501	Eastern Coach Works	H43/31F	1979	Ex Crosville, 1990
DVL641	PCA420V	Bristol VRT/SL3/501	Eastern Coach Works	H43/31F	1979	Ex Crosville, 1990
DVL642	PCA421V	Bristol VRT/SL3/501	Eastern Coach Works	H43/31F	1979	Ex Crosville, 1990
DVL643	RLG430V	Bristol VRT/SL3/501	Eastern Coach Works	H43/31F	1980	Ex Crosville, 1990
DVL644	RMA443V	Bristol VRT/SL3/501	Eastern Coach Works	H43/31F	1980	Ex Crosville, 1990
DVL645	WTU488W	Bristol VRT/SL3/501	Eastern Coach Works	H43/31F	1981	Ex Crosville, 1990
DVL646	WTU489W	Bristol VRT/SL3/501	Eastern Coach Works	H43/31F	1981	Ex Crosville, 1990
DVL647	WTU491W	Bristol VRT/SL3/501	Eastern Coach Works	H43/31F	1981	Ex Crosville, 1990
DVL667	URF667S	Bristol VRT/SL3/501	Eastern Coach Works	H43/31F	1978	
DVL674	URF674S	Bristol VRT/SL3/501	Eastern Coach Works	H43/31F	1978	
DVL685	YBF685S	Bristol VRT/SL3/501	Eastern Coach Works	H43/31F	1978	
DVL689	BRF689T	Bristol VRT/SL3/501	Eastern Coach Works	H43/31F	1978	
DVL693	BRF693T	Bristol VRT/SL3/501	Eastern Coach Works	H43/31F	1978	

DOC758, G758XRE, is painted in a silver, red and yellow livery for the 320 service from Hanley to Crewe. This service is to gain two of the new Dennis Lance which are also expected to carry this livery.

DVL701-732

Bristol VRT/SL3/501   Eastern Coach Works   H43/31F*   1979-80 *706/11/8/20/3/5/7-9/32 are DPH39/28F

701	GRF701V	709	GRF709V	715	GRF715V	720	MFA720V	728	NEH728W
704	GRF704V	710	GRF710V	716	GRF716V	723	MFA723V	729	NEH729W
706	GRF706V	711	GRF711V	717	MFA717V	725	NEH725W	731	NEH731W
707	GRF707V	714	GRF714V	718	MFA718V	727	NEH727W	732	NEH732W
708	GRF708V								

DOG733-747

Leyland Olympian ONLXB/1R   Eastern Coach Works   H45/32F   1983-84

733	A733GFA	736	A736GFA	739	A739GFA	742	A742GFA	745	A745JRE
734	A734GFA	737	A737GFA	740	A740GFA	743	A743JRE	746	A746JRE
735	A735GFA	738	A738GFA	741	A741GFA	744	A744JRE	747	A747JRE

DOG748	EWY78Y	Leyland Olympian ONLXB/1R	Roe	H47/29F	1983	Ex Turner, Brown Edge, 1988
DOG749	EWY79Y	Leyland Olympian ONLXB/1R	Roe	H47/29F	1983	Ex Turner, Brown Edge, 1988
DOG750	GFM101X	Leyland Olympian ONLXB/1R	Eastern Coach Works	H45/32F	1982	Ex Crosville, 1990
DOG751	GFM102X	Leyland Olympian ONLXB/1R	Eastern Coach Works	H45/32F	1982	Ex Crosville, 1990
DOG752	GFM103X	Leyland Olympian ONLXB/1R	Eastern Coach Works	H45/32F	1982	Ex Crosville, 1990

DOC753-762

Leyland Olympian ONCL11/1RZ   Leyland   H47/29F*   1989   *756-762 are DPH43/29F

| 753 | G753XRE | 755 | G755XRE | 757 | G757XRE | 759 | G759XRE | 761 | G761XRE |
| 754 | G754XRE | 756 | G756XRE | 758 | G758XRE | 760 | G760XRE | 762 | G762XRE |

DOG763-783

Leyland Olympian ONLXB/1R   Eastern Coach Works   H45/32F   1982-83 Ex Crosville, 1990

763	GFM104X	768	KFM111Y	772	KFM115Y	776	MTU124Y	780	A138SMA
764	GFM105X	769	KFM112Y	773	MTU120Y	777	MTU125Y	781	A143SMA
765	GFM106X	770	KFM113Y	774	MTU122Y	778	A136SMA	782	A144SMA
766	GFM108X	771	KFM114Y	775	MTU123Y	779	A137SMA	783	A145SMA
767	GFM109X								

The Leyland Lynx delivered new to PMT formed a batch of eleven fitted with high-back seating. To these have been added four older examples from Topp-Line used on the Wirral and two from Westbus, all with bus seating. Seen on service 24 is SLC861, H861GRE. *Cliff Beeton*

DOG784-799

Leyland Olympian ONLXB/1R · Eastern Coach Works · H45/32F · 1984-85 Ex Crosville, 1990

784	A146UDM	788	A159UDM	791	A162VDM	794	A165VDM	797	A168VFM
785	A156UDM	789	A160UDM	792	A163VDM	795	A166VFM	798	A169VFM
786	A157UDM	790	A161VDM	793	A164VDM	796	A167VFM	799	A170VFM
787	A158UDM								

SAD801-809

DAF SB220LC550 · Optare Delta · DP48F · 1990

801	H801GRE	803	H803GRE	805	H805GRE	807	H807GRE	809	H809GRE
802	H802GRE	804	H804GRE	806	H806GRE	808	H808GRE		

SLC845	F361YTJ	Leyland Lynx LX112L10ZR1R	Leyland Lynx	B51F	1988	Ex Topp-Line, Wavertree, 1994
SLC846	F362YTJ	Leyland Lynx LX112L10ZR1R	Leyland Lynx	B51F	1988	Ex Topp-Line, Wavertree, 1994
SLC847	F363YTJ	Leyland Lynx LX112L10ZR1R	Leyland Lynx	B51F	1988	Ex Topp-Line, Wavertree, 1994
SLC848	F364YTJ	Leyland Lynx LX112L10ZR1R	Leyland Lynx	B51F	1988	Ex Topp-Line, Wavertree, 1994
SLC849	F608WBV	Leyland Lynx LX112L10ZR1S	Leyland Lynx	B51F	1988	Ex Westbus, Ashford, 1993
SLC850	G136YRY	Leyland Lynx LX112L10ZR1R	Leyland Lynx	B51F	1990	Ex Westbus, Ashford, 1993

SLC851-861

Leyland Lynx LX2R11C15Z4S · Leyland Lynx · DP48F · 1990

851	H851GRE	854	H854GRE	856	H856GRE	858	H858GRE	860	H860GRE
852	H852GRE	855	H855GRE	857	H857GRE	859	H859GRE	861	H861GRE
853	H853GRE								

SDC862	L862HFA	Dennis Lance 11SDA3112	Northern Counties Paladin	DP47F	1993

SDC863-867

Dennis Lance 11SDA3113 · Plaxton Verde · B49F · 1995

863	N863CEH	864	N864CEH	865	N865CEH	866	N866CEH	867	N867CEH

The Pennine operation is based at a depot in Dukinfield, part of Tameside and to the east of Manchester. From here several services operate over routes out of Ashton bus station. Shown heading for Micklehurst is IDC943, M943SRE, a Dennis Dart with Marshall bodywork. *Cliff Beeton*

As this edition was being finalised the first of the Plaxton Verde-bodied Dennis Lance buses had been delivered, though had yet to enter service. Photographed the morning after delivery was SDC863, N863CEH.

DOG891	A171VFM	Leyland Olympian ONLXB/1R	Eastern Coach Works	H45/32F	1984	Ex Crosville, 1990
DOG892	B181BLG	Leyland Olympian ONLXB/1R	Eastern Coach Works	H45/32F	1984	Ex Crosville, 1990
DOG893	B182BLG	Leyland Olympian ONLXB/1R	Eastern Coach Works	H45/32F	1984	Ex Crosville, 1990
DOG894	B188BLG	Leyland Olympian ONLXB/1R	Eastern Coach Works	H45/32F	1985	Ex Crosville, 1990
DOG895	B195BLG	Leyland Olympian ONLXB/1R	Eastern Coach Works	H45/32F	1985	Ex Crosville, 1990
DOG896	B199DTU	Leyland Olympian ONLXB/1R	Eastern Coach Works	H45/32F	1985	Ex Crosville, 1990
DOG897	B200DTU	Leyland Olympian ONLXB/1R	Eastern Coach Works	DPH42/32F	1985	Ex Crosville, 1990
DOG898	B201DTU	Leyland Olympian ONLXB/1R	Eastern Coach Works	DPH42/32F	1985	Ex Crosville, 1990
DOG899	B202DTU	Leyland Olympian ONLXB/1R	Eastern Coach Works	DPH42/32F	1985	Ex Crosville, 1990
D900	WVT900S	Foden/NC 6LXB	Northern Counties	H43/31F	1978	

IDC901-920

Dennis Dart 9SDL3011 — Plaxton Pointer — DP35F — 1991-92

901	J901SEH	**905**	J905SEH	**909**	J909SEH	**913**	J913SEH	**917**	J917SEH
902	J902SEH	**906**	J906SEH	**910**	J910SEH	**914**	J914SEH	**918**	J918SEH
903	J903SEH	**907**	J907SEH	**911**	J911SEH	**915**	J915SEH	**919**	K919XRF
904	J904SEH	**908**	J908SEH	**912**	J912SEH	**916**	J916SEH	**920**	K920XRF

IDC921-929

Dennis Dart 9SDL3016 — Plaxton Pointer — DP35F — 1992

921	K921XRF	**923**	K923XRF	**925**	K925XRF	**927**	K927XRF	**929**	K929XRF
922	K922XRF	**924**	K924XRF	**926**	K926XRF	**928**	K928XRF		

IDC930-934

Dennis Dart 9SDL3034 — Plaxton Pointer — DP35F — 1993

930	L930HFA	**931**	L931HFA	**932**	L932HFA	**933**	L933HFA	**934**	L934HFA

IDC935	L935HFA	Dennis Dart 9.8SDL3025	Marshall C36	DP36F	1993
IDC936	L936HFA	Dennis Dart 9.8SDL3025	Marshall C36	DP36F	1993

IDC937-942

Dennis Dart 9SDL3034 — Plaxton Pointer — DP35F — 1994

937	L937LRF	**939**	L939LRF	**940**	L940LRF	**941**	L941LRF	**942**	L942LRF
938	L938LRF								

IDC943-952

Dennis Dart 9SDL3040 — Marshall C36 — DP35F — 1994

943	M943SRE	**945**	M945SRE	**947**	M947SRE	**949**	M949SRE	**952**	M952SRE
944	M944SRE	**946**	M946SRE	**948**	M948SRE	**951**	M951SRE		

IDC953-972

Dennis Dart 9.8SDL3054 — Plaxton Pointer — B36F — 1995

953	M953XVT	**957**	M957XVT	**961**	M961XVT	**965**	M965XVT	**969**	M969XVT
954	M954XVT	**958**	M958XVT	**962**	M962XVT	**966**	M966XVT	**970**	M970XVT
955	M955XVT	**959**	M959XVT	**963**	M963XVT	**967**	M967XVT	**971**	M971XVT
956	M956XVT	**960**	M960XVT	**964**	M964XVT	**968**	M968XVT	**972**	M972XVT

Previous Registrations:

FXI8653	B116NBF	YRF267X	GLG830X, XRF2X
XRF1X	B115NBF	XRF2X	YRE472Y

Livery: Red and yellow (Crosville, PMT, Red Rider and Pennine); brown and cream (Turners) 430/1, 550/1, 740/2, 928.

RIDER GROUP

Rider Group, Network House, Stubbs Beck Lane, West 26 Industrial Estate, Cleckheaton
BD19 4TT
Bradford Traveller, MetroChange House, 61 Hall Ings, Bradford BD1 5SQ
Calderline, Skircoat Road, Halifax HX1 2RF
Kingfisher, 8 Macaulay Street, Huddersfield HD1 2LY
Leeds Citylink, Kirkstall Garage, Kirkstall Road, Leeds LS3 1LH
Quickstep Travel, Kirkstall Garage, Kirkstall Road, Leeds LS3 1LH
Rider York, 7 James Street, York YO1 3DW

Depots: Hall Ings, Bradford; Henconner Lane, Bramley; Skircoat Road, Halifax; Leeds Road, Huddersfield; Low Road, Hunslet; Fitzwilliam Street, Kinsley; Kirkstall Road, Leeds; Torre Road, Leeds; Millwood, Todmorden and James Street

1	RYG771R	Leyland National 11351A/1R		B52F	1977	Ex West Yorkshire, 1989
3	WUM443S	Leyland National 11351A/1R		B52F	1977	Ex West Yorkshire, 1989
6	PEV706R	Leyland National 11351A/1R		B49F	1977	Ex Wilts & Dorset, 1992
10	YWW810S	Leyland National 11351A/1R		B52F	1978	Ex West Yorkshire, 1989
16	L516EHD	DAF SB220LC550	Ikarus CitiBus	B48F	1993	
17	L517EHD	DAF SB220LC550	Ikarus CitiBus	B48F	1994	
18	L518EHD	DAF SB220LC550	Ikarus CitiBus	B48F	1994	
23	UOI4323	Volvo B10M-61	East Lancashire (1993)	B51F	1982	Ex Rhodeservices, 1994
26	K506RJX	DAF SB220LC550	Ikarus CitiBus	B48F	1993	Ex Yorkshire Travel, 1994
27	K507RJX	DAF SB220LC550	Ikarus CitiBus	B48F	1993	Ex Yorkshire Travel, 1994
29	F229FSU	Leyland Tiger TRBTL11/2RP	Plaxton Derwent II	B54F	1988	Ex Rhodeservices, 1994
30	F300GNS	Leyland Tiger TRBTL11/2RP	Plaxton Derwent II	B54F	1988	Ex Rhodeservices, 1994
42	TYG742R	Leyland National 11351A/1R		B52F	1977	Ex West Yorkshire, 1989
43	JUB643V	Leyland National 11351A/1R		B52F	1979	Ex West Yorkshire, 1989
44	JUB644V	Leyland National 11351A/1R		B52F	1979	Ex West Yorkshire, 1989
45	E345SWY	Iveco Daily 49-10	Robin Hood City Nippy	B25F	1988	Ex West Yorkshire, 1989
51	MEL551P	Leyland National 11351/1R		B49F	1976	Ex Wilts & Dorset, 1993
66	FPR66V	Leyland National 11351A/1R		B49F	1980	Ex Wilts & Dorset, 1992
71	E471TYG	Iveco Daily 49-10	Robin Hood City Nippy	B19F	1988	Ex West Yorkshire, 1989

228-236

		Renault-Dodge S56	Reeve Burgess Beaver	B23F	1989	Ex York City & District, 1990

228	G251LWF	230	G253LWF	232	G255LWF	234	G257LWF	236	G259LWF
229	G252LWF	231	G254LWF	233	G256LWF	235	G258LWF		

238	G447LKW	Renault-Dodge S56	Reeve Burgess Beaver	B23F	1990	Ex York City & District, 1990
290	E166CNC	Renault-Dodge S56	Northern Counties	B23F	1988	Ex Reynard Buses, 1990
291	E385CNE	Renault-Dodge S56	Northern Counties	B23F	1988	Ex Reynard Buses, 1990

332	MNW132V	Leyland National 2 NL116L11/1R	B52F	1980	Ex York City & District, 1990
333	SWX533W	Leyland National 2 NL116AL11/1R	B52F	1981	Ex York City & District, 1990
335	SWX535W	Leyland National 2 NL116AL11/1R	B52F	1981	Ex York City & District, 1990
338	SWX538W	Leyland National 2 NL116AL11/1R	B52F	1981	Ex York City & District, 1990
340	SWX540W	Leyland National 2 NL116AL11/1R	B52F	1981	Ex York City & District, 1990
363	PNW603W	Leyland National 2 NL116L11/1R	B52F	1980	Ex York City & District, 1990
368	UWY68X	Leyland National 2 NL116AL11/1R	B52F	1982	Ex York City & District, 1990
371	UWY71X	Leyland National 2 NL116AL11/1R	B52F	1982	Ex York City & District, 1990

713-756

		Bristol VRT/SL3/6LXB	Eastern Coach Works	H43/31F	1978-81	Ex West Yorkshire, 1989

713	DWU296T	729	JWT762V	744	LWU472V	748	NUM339V	755	PWY41W
717	EWR165T	735	LUA718V	745	PUM148W	752	PWY38W	756	PWY42W
725	JWT758V	736	LUA719V	746	PUM149W				

Opposite, top: **One of the large batch of Plaxton-bodied Dennis Lances delivered in 1995 4025, M425VWW is seen in Keighley. Similar 4041-4046 received 'N' marks their later arrival being due to their differing specification for Greater Manchester tendered service.**
Opposite, bottom: **Seen route testing prior to the main launch of the York park & ride contract is 8412, N412ENW, one of the first Wright Axcess-bodied Scania L113CRLs. The fourteen delivered in 1995 will be boosted by a further six in 1996 and will link three peripheral city car parks with the centre. Temporary vinyl destinations are now replaced by blinds.** *Andrew Jarosz*

980-994

		Bristol VRT/SL3/6LXB		Eastern Coach Works	H43/31F	1977-78 Ex West Yorkshire, 1989

980	SWW305R	988	WWY118S	992	WWY122S	993	WWY123S	994	WWY127S

1001-1030

Volvo B10B-58 — Alexander Strider — B51F — 1993-94

1001	K101HUM	1007	K107HUM	1013	K113HUM	1019	K119HUM	1025	L125PWR
1002	K102HUM	1008	K108HUM	1014	K114HUM	1020	K120HUM	1026	L126PWR
1003	K103HUM	1009	K109HUM	1015	K115HUM	1021	L121PWR	1027	L127PWR
1004	K104HUM	1010	K110HUM	1016	K116HUM	1022	L122PWR	1028	L128PWR
1005	K105HUM	1011	K211HUM	1017	K117HUM	1023	L123PWR	1029	L129PWR
1006	K106HUM	1012	K112HUM	1018	K118HUM	1024	L124PWR	1030	L130PWR

1171	B124PEL	Bedford YNT	Plaxton Paramount 3200 E	C53F	1984	Ex Reynard Buses, 1990

1201-1208

DAF SB220LC550 — Ikarus CitiBus — B48F — 1992

1201	J421NCP	1203	J423NCP	1205	J425NCP	1207	K527RJX	1208	K528RJX
1202	J422NCP	1204	J424NCP	1206	J426NCP				

1251-1255

DAF SB220LC550 — Optare Delta — B47F — 1989

1251	G251JYG	1252	G252JYG	1253	G253JYG	1254	G254JYG	1255	G255JYG

1256	J994GCP	DAF SB220LC550	Optare Delta	B49F	1991	Ex Yorkshire Travel, 1994

1317-1329

Leyland National 2 NL106L11/1R — B41F — 1980 Ex WYPTE, 1986

1317	LUA317V	1319	LUA319V	1321	LUA321V	1323	LUA323V	1329	LUA329V
1318	LUA318V	1320	LUA320V	1322	LUA322V				

1331	VWU331X	Leyland National 2 NL116AL11/1R	B49F	1981	Ex WYPTE, 1986
1332	VWU332X	Leyland National 2 NL116AL11/1R	B49F	1981	Ex WYPTE, 1986
1333	YWX333X	Leyland National 2 NL116AL11/1R	B49F	1982	Ex WYPTE, 1986
1339	JUB645V	Leyland National 11351A/1R	B52F	1979	Ex West Yorkshire, 1989
1340	JUB646V	Leyland National 11351A/1R(Volvo)	B49F	1979	Ex West Yorkshire, 1989
1341	JUB647V	Leyland National 11351A/1R(Volvo)	B49F	1979	Ex West Yorkshire, 1989
1342	JUB649V	Leyland National 11351A/1R	B52F	1979	Ex West Yorkshire, 1989
1343	MNW130V	Leyland National 2 NL116L11/1R	B52F	1980	Ex West Yorkshire, 1989
1345	PNW598W	Leyland National 2 NL116L11/1R	B49F	1980	Ex West Yorkshire, 1989
1348	PNW601W	Leyland National 2 NL116L11/1R	B49F	1980	Ex West Yorkshire, 1989

1349-1354

Leyland National 2 NL116AL11/1R — B52F — 1981-82 Ex West Yorkshire, 1989

1349	SWX537W	1351	UWY72X	1352	UWY74X	1353	UWY75X	1354	UWY90X
1350	UWY69X								

1355	PWY587W	Leyland National 2 NL106AL11/1R		B44F	1981	Ex West Yorkshire, 1989
1356	PWY588W	Leyland National 2 NL106AL11/1R		B44F	1981	Ex West Yorkshire, 1989
1357	FWA473V	Leyland National 2 NL106AL11/1R		B41F	1980	Ex Pride of the Road, 1993
1366	RKA869T	Leyland National 11351A/1R(Volvo)		B49F	1978	Ex Pride of the Road, 1993
1367	MHJ723V	Leyland National 2 NL116AL11/1R		B49F	1980	Ex Eastern National, 1994
1368	MHJ726V	Leyland National 2 NL116AL11/1R		B49F	1980	Ex Eastern National, 1994
1369	MHJ729V	Leyland National 2 NL116AL11/1R		B49F	1980	Ex Eastern National, 1994
1370	STW19W	Leyland National 2 NL116AL11/1R		B49F	1980	Ex Eastern National, 1994
1400	YR3939	Volvo B10M-61	Jonckheere Jubilee P599	C32FT	1988	
1402	NIB4906	Volvo B10M-61	Jonckheere Jubilee P50	C53F	1986	Ex WYPTE, 1986
1403	NIB4905	Volvo B10M-61	Jonckheere Jubilee P50	C48FT	1987	
1404	NIB4908	Volvo B10M-61	Jonckheere Jubilee P50	C48FT	1987	
1405	GSU388	Volvo B10M-61	Jonckheere Jubilee P599	C51FT	1987	
1406	E406RWR	Volvo B10M-61	Duple 340	C57F	1988	
1407	GSU390	Volvo B10M-61	Duple 340	C57F	1988	
1416	D783SGB	Volvo B10M-61	Plaxton Paramount 3500 III	C53FT	1987	Ex Park's, 1988
1418	23PTA	Volvo B10M-60	Jonckheere Deauville P599	C48FT	1989	
1419	8995WY	Volvo B10M-60	Plaxton Paramount 3500 III	C49FT	1990	Ex Park's, 1992
1420	G76RGG	Volvo B10M-60	Plaxton Paramount 3500 III	C49FT	1990	Ex Park's, 1992
1421	H841AHS	Volvo B10M-60	Plaxton Paramount 3500 III	C53F	1991	Ex Park's, 1993
1422	L22YRL	Volvo B10M-60	Jonckheere Deauville P599	C51F	1993	
1423	L511NYG	Volvo B10M-60	Plaxton Première 350	C49FT	1993	
1424	L541XUT	Volvo B10M-60	Plaxton Première 350	C49FT	1994	
1425	L542XUT	Volvo B10M-60	Plaxton Première 350	C49FT	1994	
1426	L546XUT	Volvo B10M-62	Plaxton Première 350	C49FT	1994	

The low-cost, Leeds-based unit, Quickstep, took over some Rhodeservice routes in 1994 and received a number of Iveco Daily minibuses which had originally operated for West Yorkshire Road Car but latterly running in Bradford. A recent victim of withdrawal was 64, E464TYG.
Phillip Stephenson

Originally delivered to Yorkshire Rider in Huddersfield in 1993, the first batch of Volvo B10Bs with Alexander Strider bodies was transferred to Halifax during the summer of 1995. One of the first vehicles to carry the new Calderline fleet names, 1012, K112HUM, revisits its old home on a limited stop service.
Andrew Jarosz

Yorkshire Rider switched to Jonckheere in 1986 when it purchased its first Volvo B10M coaches, after buying Plaxton-bodied Tigers in the early 1980s. The Gold Rider high-quality coaching operation has been reduced in recent times although 1402, NIB4906 remains.
Colin Lloyd

1451-1460

Volvo B10M-55 — Plaxton Derwent II — B51F — 1990

1451	G451JYG	**1453**	G453JYG	**1455**	G455JYG	**1457**	G457JYG	**1459**	G459JYG
1452	G452JYG	**1454**	G454JYG	**1456**	G456JYG	**1458**	G458JYG	**1460**	G460JYG

1528	GWU528T	Leyland Leopard PSU3E/4R	Plaxton Supreme IV Express	C53F	1979	Ex WYPTE, 1986
1532	JUM532V	Leyland Leopard PSU3E/4R	Plaxton Supreme IV Express	C53F	1979	Ex WYPTE, 1986
1544	KUB548V	Leyland Leopard PSU3C/4R	Plaxton Supreme IV Express	C49F	1980	Ex West Yorkshire, 1989
1547	NPA226W	Leyland Leopard PSU3F/4R	Duple Dominant IV Express	C49F	1981	Ex West Yorkshire, 1989
1553	UWY86X	Leyland Leopard PSU3F/4R	Duple Dominant IV Express	C49F	1981	Ex West Yorkshire, 1989
1581	NPA226W	Leyland Leopard PSU3E/4R	Plaxton Supreme IV Express	C53F	1981	Ex Reynard Buses, 1990
1587	CSU244	Leyland Leopard PSU3D/4R	Plaxton Bustler (1987)	B55F	1977	Ex Reynard Buses, 1990
1592	LOI6690	Leyland Leopard PSU3D/4R	Plaxton Derwent (1987)	B55F	1977	Ex Reynard Buses, 1990
1593	VDH244S	Leyland Leopard PSU3E/4R	Duple Dominant (1985)	B55F	1977	Ex Reynard Buses, 1990
1594	ABR868S	Leyland Leopard PSU3E/4R	Plaxton Derwent II (1990)	B55F	1977	Ex Reynard Buses, 1990
1595	JKW215W	Leyland Leopard PSU3G/4R	Plaxton Derwent II (1990)	B55F	1981	Ex Reynard Buses, 1990
1596	JKW216W	Leyland Leopard PSU3G/4R	Plaxton Derwent II (1990)	B55F	1981	Ex Reynard Buses, 1990
1597	ABR869S	Leyland Leopard PSU3E/4R	Plaxton Derwent II (1990)	B55F	1977	Ex Reynard Buses, 1990
1604	WSV408	Leyland Tiger TRCTL11/3R	Plaxton Paramount 3200	C53F	1983	Ex WYPTE, 1986
1605	WSV409	Leyland Tiger TRCTL11/3R	Plaxton Paramount 3200	C53F	1983	Ex WYPTE, 1986
1606	HUA606Y	Leyland Tiger TRCTL11/2RH	Plaxton Paramount 3200 E	C49F	1983	Ex WYPTE, 1986
1607	HUA607Y	Leyland Tiger TRCTL11/2RH	Plaxton Paramount 3200 E	C49F	1983	Ex WYPTE, 1986
1608	WSV410	Leyland Tiger TRCTL11/3R	Plaxton Paramount 3200	C49F	1984	Ex WYPTE, 1986
1610	B610VWU	Leyland Tiger TRCTL11/3RH	Plaxton Paramount 3200 IIE	C53F	1985	Ex WYPTE, 1986

1615-1622

Leyland Tiger TRCTL11/3ARZA — Plaxton Paramount 3200 IIIE — C53F — 1988 — *1619/20/2 are Paramount 3200 III

1615	F615XWY	**1617**	F617XWY	**1619**	F619XWY	**1621**	F621XWY	**1622**	F622XWY
1616	F616XWY	**1618**	F618XWY	**1620**	F620XWY				

1623	EWW945Y	Leyland Tiger TRCTL11/3R	Plaxton Paramount 3200 E	C53F	1983	Ex West Yorkshire, 1989

1654-1672

Leyland Tiger TRBTL11/2R — Duple Dominant — DP47F — 1983-84 Ex WYPTE, 1986

1654	EWR654Y	**1657**	EWR657Y	**1661**	A661KUM	**1666**	A666KUM	**1670**	A670KUM
1655	EWR655Y	**1658**	A658KUM	**1662**	A662KUM	**1667**	A667KUM	**1671**	A671KUM
1656	EWR656Y	**1659**	A659KUM	**1664**	A664KUM	**1669**	A669KUM	**1672**	A672KUM

1793-1874

Freight Rover Sherpa — Dormobile — B20F — 1987

1793	D793KWR	**1799**	D799KWR	**1865**	D865LWR	**1869**	D869LWR	**1871**	D871LWR
1796	D796KWR	**1862**	D862LWR	**1868**	D868LWR	**1870**	D870LWR	**1874**	D874LWR

2001	E201PWY	Mercedes-Benz 811D	Optare StarRider	DP33F	1987	
2002	E202PWY	Mercedes-Benz 811D	Optare StarRider	DP33F	1987	
2003	E203PWY	Mercedes-Benz 811D	Optare StarRider	DP29F	1987	
2004	E204PWY	Mercedes-Benz 811D	Optare StarRider	DP29F	1987	
2005	F546EJA	Mercedes-Benz 709D	PMT	DP25F	1988	Ex Rhodeservice, 1994

2006-2011

Mercedes-Benz L608D — Dormobile — B20F — 1986 Ex City Line, 1994

2006	D535FAE	**2008**	D537FAE	**2009**	D541FAE	**2010**	D539FAE	**2011**	D540FAE
2007	D536FAE								

2012	C222HJN	Mercedes-Benz L608D	Reeve Burgess	B20F	1985	Ex City Line, 1994
2013	D529FAE	Mercedes-Benz L608D	Dormobile	B20F	1986	Ex City Line, 1994
2014	D530FAE	Mercedes-Benz L608D	Dormobile	B20F	1986	Ex City Line, 1994
2015	D531FAE	Mercedes-Benz L608D	Dormobile	B20F	1986	Ex City Line, 1994
2016	D532FAE	Mercedes-Benz L608D	Dormobile	B20F	1986	Ex City Line, 1994

2067-2081

Iveco Daily 49-10 — Robin Hood City Nippy — B23F* — 1988 — Ex West Yorkshire, 1989 — *2070 is B25F

2067	E326SWY	**2071**	E331SWY	**2076**	E337SWY	**2078**	E339SWY	**2081**	E342SWY
2070	E330SWY	**2075**	E336SWY						

2093	E467TYG	Iveco Daily 49-10	Robin Hood City Nippy	B21F	1988	Ex West Yorkshire, 1989

Kingfisher is the brandname of the Huddersfield-based fleet, and an attractive motif would have gone well with the bare fleetname. Kingfisher is planning to retain the new Rider livery introduced in 1995 as 110 of its new buses already wear these colours. The first vehicle to carry the new name 2227, M227VWU is one of forty-four Plaxton bodied Mercedes-Benz 709Ds delivered this year. The major restructuring of the Rider Group was not complete in all details as this book went to press.
Andrew Jarosz

2100-2107

Iveco Daily 49-10   Phoenix   B19F   1989   Ex York City & District, 1990

2100	G210KUA	2102	G212KUA	2104	G214KUA	2106	G216KUA	2107	G217KUA
2101	G211KUA	2103	G213KUA	2105	G215KUA				

2110-2120

Mercedes-Benz L608D   Dormobile   B20F   1986   Ex City Line, 1994

2110	D554FAE	2113	D560FAE	2115	D562FAE	2117	D564FAE	2119	D567FAE
2111	D555FAE	2114	D561FAE	2116	D563FAE	2118	D566FAE	2120	D568FAE
2112	D556FAE								

2201-2263

Mercedes-Benz 709D   Plaxton Beaver   B25F   1994-95

2201	M201VWU	2214	M214VWU	2227	M227VWU	2240	M240VWU	2252	M252VWU
2202	M202VWU	2215	M215VWU	2228	M228VWU	2241	M241VWU	2253	M253VWU
2203	M203VWU	2216	M216VWU	2229	M229VWU	2242	M242VWU	2254	M254VWU
2204	M204VWU	2217	M217VWU	2230	M230VWU	2243	M243VWU	2255	M255VWU
2205	M205VWU	2218	M218VWU	2231	M231VWU	2244	M244VWU	2256	M256VWU
2206	M206VWU	2219	M219VWU	2232	M232VWU	2245	M245VWU	2257	M257VWU
2207	M207VWU	2220	M220VWU	2233	M233VWU	2246	M246VWU	2258	M258VWU
2208	M208VWU	2221	M221VWU	2234	M234VWU	2247	M247VWU	2259	M259VWU
2209	M209VWU	2222	M449VWW	2235	M235VWU	2248	M248VWU	2260	M260VWU
2210	M210VWU	2223	M223VWU	2236	M236VWU	2249	M249VWU	2261	M261VWU
2211	M211VWU	2224	M224VWU	2237	M237VWU	2250	M250VWU	2262	M262VWU
2212	M212VWU	2225	M225VWU	2238	M238VWU	2251	M251VWU	2263	M263VWU
2213	M213VWU	2226	M226VWU	2239	M239VWU				

3001-3006 — Volvo B6-9.8m — Alexander Dash — B40F — 1994

3001	L101PWR	3003	L103PWR	3004	L104PWR	3005	L105PWR	3006	L106PWR
3002	L102PWR								

3201-3218 — Dennis Dart 9.8SDL3054 — Plaxton Pointer — B40F — 1995

3201	M201VWW	3205	M205VWW	3209	M209VWW	3213	M213VWW	3216	M216VWW
3202	M202VWW	3206	M206VWW	3210	M210VWW	3214	M214VWW	3217	M217VWW
3203	M203VWW	3207	M207VWW	3211	M211VWW	3215	M215VWW	3218	M218VWW
3204	M204VWW	3208	M208VWW	3212	M212VWW				

3219-3268 — Dennis Dart 9.8SDL3054 — Alexander Dash — B40F — 1995

3219	M219VWW	3229	M229VWW	3239	M239VWW	3249	M249VWW	3259	M259VWW
3220	M220VWW	3230	M230VWW	3240	M240VWW	3250	M250VWW	3260	M260VWW
3221	M221VWW	3231	M231VWW	3241	M241VWW	3251	M251VWW	3261	M261VWW
3222	M450VWW	3232	M232VWW	3242	M242VWW	3252	M252VWW	3262	M262VWW
3223	M223VWW	3233	M233VWW	3243	M243VWW	3253	M253VWW	3263	M263VWW
3224	M224VWW	3234	M234VWW	3244	M244VWW	3254	M254VWW	3264	M264VWW
3225	M225VWW	3235	M235VWW	3245	M245VWW	3255	M255VWW	3265	M265VWW
3226	M226VWW	3236	M236VWW	3246	M246VWW	3256	M256VWW	3266	M266VWW
3227	M227VWW	3237	M237VWW	3247	M247VWW	3257	M257VWW	3267	M267VWW
3228	M228VWW	3238	M238VWW	3248	M248VWW	3258	M258VWW	3268	M268VWW

4001-4048 — Dennis Lance 11SDA3113 — Plaxton Verde — B49F — 1995

4001	M401VWW	4011	M411VWW	4021	M421VWW	4031	M431VWW	4040	M440VWW
4002	M402VWW	4012	M412VWW	4022	M422VWW	4032	M432VWW	4041	N441ENW
4003	M403VWW	4013	M413VWW	4023	M423VWW	4033	M433VWW	4042	N442ENW
4004	M404VWW	4014	M414VWW	4024	M424VWW	4034	M434VWW	4043	N443ENW
4005	M405VWW	4015	M415VWW	4025	M425VWW	4035	M435VWW	4044	N544ENW
4006	M406VWW	4016	M416VWW	4026	M426VWW	4036	M436VWW	4045	N445ENW
4007	M407VWW	4017	M417VWW	4027	M427VWW	4037	M437VWW	4046	N446ENW
4008	M408VWW	4018	M418VWW	4028	M428VWW	4038	M438VWW	4047	M447VWW
4009	M409VWW	4019	M419VWW	4029	M429VWW	4039	M439VWW	4048	M448VWW
4010	M410VWW	4020	M420VWW	4030	M430VWW				

5018-5081 — Leyland Olympian ONLXB/1R — Roe — H47/29F — 1982-83 Ex WYPTE, 1986

5018	UWW18X	5026	CUB26Y	5034	CUB34Y	5042	CUB42Y	5052	CUB52Y
5019	UWW19X	5027	CUB27Y	5035	CUB35Y	5043	CUB43Y	5053	CUB53Y
5020	UWW20X	5028	CUB28Y	5036	CUB36Y	5044	CUB44Y	5054	CUB54Y
5021	CUB21Y	5029	CUB29Y	5037	CUB37Y	5045	CUB45Y	5055	CUB55Y
5022	CUB22Y	5030	CUB30Y	5038	CUB38Y	5046	CUB46Y	5065	CUB65Y
5023	CUB23Y	5031	CUB31Y	5039	CUB39Y	5047	CUB47Y	5077	EWY77Y
5024	CUB24Y	5032	CUB32Y	5040	CUB40Y	5048	CUB48Y	5081	EWY81Y
5025	CUB25Y	5033	CUB33Y	5041	CUB41Y	5051	CUB51Y		

5082-5121 — Leyland Olympian ONLXB/1R — Roe — H47/29F — 1983 Ex WYPTE, 1986

5082	A82KUM	5090	A90KUM	5098	A98KUM	5106	A106KUM	5114	A114KUM
5083	A83KUM	5091	A91KUM	5099	A99KUM	5107	A107KUM	5115	A115KUM
5084	A84KUM	5092	A92KUM	5100	A100KUM	5108	A108KUM	5116	A116KUM
5085	A85KUM	5093	A93KUM	5101	A101KUM	5109	A109KUM	5117	A117KUM
5086	A86KUM	5094	A94KUM	5102	A102KUM	5110	A110KUM	5118	A118KUM
5087	A87KUM	5095	A95KUM	5103	A103KUM	5111	A111KUM	5119	A119KUM
5088	A88KUM	5096	A96KUM	5104	A104KUM	5112	A112KUM	5120	A120KUM
5089	A89KUM	5097	A97KUM	5105	A105KUM	5113	A113KUM	5121	A121KUM

5122-5145 — Leyland Olympian ONLXB/1R — Roe — H47/29F — 1984 Ex WYPTE, 1986

5122	B122RWY	5127	B127RWY	5132	B132RWY	5137	B137RWY	5142	B503RWY
5123	B123RWY	5128	B128RWY	5133	B133RWY	5138	B138RWY	5143	B504RWY
5124	B124RWY	5129	B129RWY	5134	B134RWY	5139	B139RWY	5144	B505RWY
5125	B125RWY	5130	B130RWY	5135	B135RWY	5140	B501RWY	5145	B140RWY
5126	B126RWY	5131	B131RWY	5136	B136RWY	5141	B502RWY		

Numerically the last of the Plaxton-bodied Dennis Lances delivered to Huddersfield in 1995, 4048 M448VWW was delivered out of sequence in order to receive a full-rear advertisement for a local car dealer. As mentioned earlier, lower numbered vehicles received 'N' marks. *Andrew Jarosz*

5146-5150

Leyland Olympian ONLXB/1R		Optare			H47/29F*		1985		*5146/7 are CO47/29F Ex WYPTE, 1986

5146	C146KBT	5147	C147KBT	5148	C148KBT	5149	C149KBT	5150	C150KBT

5151-5175

Leyland Olympian ONCL10/1RZ Northern Counties   H45/29F   1988

5151	F151XYG	5156	F156XYG	5161	F161XYG	5166	F166XYG	5171	F171XYG
5152	F152XYG	5157	F157XYG	5162	F162XYG	5167	F167XYG	5172	F172XYG
5153	F153XYG	5158	F158XYG	5163	F163XYG	5168	F168XYG	5173	F173XYG
5154	F154XYG	5159	F159XYG	5164	F164XYG	5169	F169XYG	5174	F174XYG
5155	F155XYG	5160	F160XYG	5165	F165XYG	5170	F170XYG	5175	F175XYG

5176-5185

Leyland Olympian ONCL10/1RZ Northern Counties   H47/29F   1990

5176	G176JYG	5178	G178JYG	5180	G180JYG	5182	G182JYG	5184	G184JYG
5177	G177JYG	5179	G179JYG	5181	G181JYG	5183	G183JYG	5185	G185JYG

5186-5199

Leyland Olympian ONLXB/1R   Eastern Coach Works   H45/32F   1983-85 Ex West Yorkshire, 1989

5186	FUM486Y	5189	FUM492Y	5192	FUM498Y	5195	A599NYG	5198	C483YWY
5187	FUM487Y	5190	FUM494Y	5193	FUM499Y	5196	A600NYG	5199	C485YWY
5188	FUM491Y	5191	FUM495Y	5194	A686MWX	5197	A601NYG		

5200-5222

Leyland Olympian ONCL10/1RZ Alexander RL   H47/32F*   1990   *5207-13 are H47/30F

5200	G623OWR	5205	G605OWR	5210	G610OWR	5215	G615OWR	5219	G619OWR
5201	G601OWR	5206	G606OWR	5211	G611OWR	5216	G616OWR	5220	G620OWR
5202	G602OWR	5207	G607OWR	5212	G612OWR	5217	G617OWR	5221	G621OWR
5203	G603OWR	5208	G608OWR	5213	G613OWR	5218	G618OWR	5222	G622OWR
5204	G604OWR	5209	G609OWR	5214	G614OWR					

5301-5315

Volvo Olympian YN2RV18Z4 · Northern Counties Palatine · H47/29F · 1994

5301	L301PWR	5304	L304PWR	5307	L307PWR	5310	L310PWR	5313	L313PWR
5302	L302PWR	5305	L305PWR	5308	L308PWR	5311	L311PWR	5314	L314PWR
5303	L303PWR	5306	L306PWR	5309	L309PWR	5312	L312PWR	5315	L315PWR

5401-5405

Volvo Olympian YN2RC16Z4 · Northern Counties Palatine · H47/29F · 1994

5401	L401PWR	5402	L402PWR	5403	L403PWR	5404	L404PWR	5405	L405PWR

5501-5506

Leyland Olympian ONLXB/1R · Roe · H47/27F · 1984 · Ex WYPTE, 1986

5501	B141RWY	5503	B143RWY	5504	B144RWY	5505	B145RWY	5506	B506RWY
5502	B142RWY								

5507-5511

Leyland Olympian ONTL11/1R · Optare · H47/27F · 1985 · Ex WYPTE, 1986

5507	C507KBT	5508	C508KBT	5509	C509KBT	5510	C510KBT	5511	C511KBT

5512-5516

Leyland Olympian ONTL11/1R · Optare · DPH43/27F* 1987 · *5512/4 are H47/27F

5512	D512HUB	5513	D513HUB	5514	D514HUB	5515	D515HUB	5516	D516HUB

5517	FUM489Y	Leyland Olympian ONLXB/1R	Eastern Coach Works	H41/29F	1983	Ex West Yorkshire, 1989
5518	B518UWW	Leyland Olympian ONLXB/1R	Eastern Coach Works	DPH42/29F	1985	Ex West Yorkshire, 1989
5519	B519UWW	Leyland Olympian ONLXB/1R	Eastern Coach Works	DPH42/29F	1985	Ex West Yorkshire, 1989
5520	B520UWW	Leyland Olympian ONLXB/1R	Eastern Coach Works	DPH42/29F	1985	Ex West Yorkshire, 1989
5521	B521UWW	Leyland Olympian ONLXB/1R	Eastern Coach Works	DPH42/29F	1985	Ex West Yorkshire, 1989

5601-5605

Volvo Olympian YN2RV18Z4 · Northern Counties Palatine · DPH47/29F 1994

5601	L601PWR	5602	L602PWR	5603	L603PWR	5604	L604PWR	5605	L605PWR

6001-6074

Leyland Atlantean AN68/1R · Roe · H43/33F · 1974-75 Ex WYPTE, 1986

6001	GUG533N	6027	GUG554N	6038	GUG565N	6045	HWT31N	6064	HWT50N
6005	GUG535N	6030	GUG557N	6039	GUG566N	6047	HWT33N	6068	HWT54N
6015	GUG542N	6032	GUG559N	6040	GUG567N	6050	HWT36N	6073	HWT59N
6017	GUG544N	6033	GUG560N	6043	HWT29N	6058	HWT44N	6074	HWT60N
6024	GUG551N	6037	GUG564N	6044	HWT30N	6061	HWT47N		

6113-6175

Leyland Atlantean AN68/1R · Roe · H43/33F · 1975-77 Ex WYPTE, 1986

6113	LUG113P	6132	SUA132R	6147	SUA147R	6159	WNW159S	6167	WNW167S
6116	LUG116P	6133	SUA133R	6150	SUA150R	6161	WNW161S	6169	WNW169S
6117	LUG117P	6134	SUA134R	6152	WNW152S	6162	WNW162S	6171	WNW171S
6118	LUG118P	6140	SUA140R	6156	WNW156S	6163	WNW163S	6172	WNW172S
6127	SUA127R	6142	SUA142R	6157	WNW157S	6164	WNW164S	6174	WNW174S
6128	SUA128R	6144	SUA144R	6158	WNW158S	6165	WNW165S	6175	WNW175S
6130	SUA130R	6146	SUA146R						

Representing a modest influx of twenty-five double deck buses to Yorkshire Rider in 1994, Northern Counties Palatine-bodied Volvo Olympian 5403, L403PWR was one of ten then allocated to Bradford.
Malc McDonald

6176-6191
Leyland Atlantean AN68/1R   Roe   H43/33F   1979   Ex WYPTE, 1986

6176	GWR176T	6180	GWR180T	6183	GWR183T	6187	GWR187T	6190	GWR190T
6177	GWR177T	6181	GWR181T	6185	GWR185T	6188	GWR188T	6191	GWR191T
6178	GWR178T	6182	GWR182T	6186	GWR186T	6189	GWR189T		

6192-6266
Leyland Atlantean AN68A/1R   Roe   H43/32F   1979-80 Ex WYPTE, 1986

6192	JUM192V	6207	JUM207V	6222	KWY222V	6238	KWY238V	6253	PUA253W
6193	JUM193V	6208	JUM208V	6223	KWY223V	6239	KWY239V	6254	PUA254W
6194	JUM194V	6209	JUM209V	6224	KWY224V	6241	KWY241V	6255	PUA255W
6195	JUM195V	6210	JUM210V	6225	KWY225V	6242	KWY242V	6256	PUA256W
6196	JUM196V	6211	JUM211V	6226	KWY226V	6243	KWY243V	6257	PUA257W
6197	JUM197V	6212	JUM212V	6227	KWY227V	6244	KWY244V	6258	PUA258W
6198	JUM198V	6213	JUM213V	6228	KWY228V	6245	KWY245V	6259	PUA259W
6199	JUM199V	6214	JUM214V	6229	KWY229V	6246	KWY246V	6260	PUA260W
6200	JUM200V	6215	JUM215V	6230	KWY230V	6247	KWY247V	6261	PUA261W
6201	JUM201V	6216	KWY216V	6231	KWY231V	6248	KWY248V	6262	PUA262W
6202	JUM202V	6217	KWY217V	6232	KWY232V	6249	KWY249V	6263	PUA263W
6203	JUM203V	6218	KWY218V	6233	KWY233V	6250	KWY250V	6264	PUA264W
6204	JUM204V	6219	KWY219V	6234	KWY234V	6251	KWY251V	6265	PUA265W
6205	JUM205V	6220	KWY220V	6236	KWY236V	6252	PUA252W	6266	PUA266W
6206	JUM206V	6221	KWY221V	6237	KWY237V				

6267-6326
Leyland Atlantean AN68C/1R   Roe   H43/32F   1980-81 Ex WYPTE, 1986

6267	PUA267W	6279	PUA279W	6292	PUA292W	6304	PUA304W	6316	PUA316W
6268	PUA268W	6280	PUA280W	6293	PUA293W	6305	PUA305W	6317	PUA317W
6269	PUA269W	6282	PUA282W	6294	PUA294W	6306	PUA306W	6318	PUA318W
6270	PUA270W	6283	PUA283W	6295	PUA295W	6307	PUA307W	6319	PUA319W
6271	PUA271W	6284	PUA284W	6296	PUA296W	6308	PUA308W	6320	PUA320W
6272	PUA272W	6285	PUA285W	6297	PUA297W	6309	PUA309W	6321	PUA321W
6273	PUA273W	6286	PUA286W	6298	PUA298W	6310	PUA310W	6322	PUA322W
6274	PUA274W	6287	PUA287W	6299	PUA299W	6311	PUA311W	6323	PUA323W
6275	PUA275W	6288	PUA288W	6300	PUA300W	6312	PUA312W	6324	PUA324W
6276	PUA276W	6289	PUA289W	6301	PUA301W	6313	PUA313W	6325	PUA325W
6277	PUA277W	6290	PUA290W	6302	PUA302W	6314	PUA314W	6326	PUA326W
6278	PUA278W	6291	PUA291W	6303	PUA303W				

6327-6361
Leyland Atlantean AN68C/1R   Roe   H43/32F   1981   Ex WYPTE, 1986

6327	VWW327X	6334	VWW334X	6341	VWW341X	6348	VWW348X	6355	VWW355X
6328	VWW328X	6335	VWW335X	6342	VWW342X	6349	VWW349X	6356	VWW356X
6329	VWW329X	6336	VWW336X	6343	VWW343X	6350	VWW350X	6357	VWW357X
6330	VWW330X	6337	VWW337X	6344	VWW344X	6351	VWW351X	6358	VWW358X
6331	VWW331X	6338	VWW338X	6345	VWW345X	6352	VWW352X	6359	VWW359X
6332	VWW332X	6339	VWW339X	6346	VWW346X	6353	VWW353X	6360	VWW360X
6333	VWW333X	6340	VWW340X	6347	VWW347X	6354	VWW354X	6361	VWW361X

Showing how an Atlantean would look in Kingfisher colours, 6338 VWW338X was the first and so far the only AN68 to wear them. Seen on service in Huddersfield during September 1995. The vehicle has now been rebranded with special advertising for the Huddersfield University shuttle.
Andrew Jarosz

6423	UNA800S	Leyland Atlantean AN68/1R	Northern Counties	H43/32F	1978	Ex G M Buses, 1988

6425-6435

		Leyland Atlantean AN68B/1R	Roe		H43/30F	1980-81 Ex Sovereign, 1989

6425	EPH227V	6428	KPJ257W	6430	KPJ261W	6432	KPJ290W	6434	KPJ292W
6426	KPJ253W	6429	KPJ260W	6431	KPJ263W	6433	KPJ291W	6435	MPG293W
6427	KPJ255W								

7030	MNW30P	Leyland Fleetline FE30GR	Roe	H43/33F	1976	Ex WYPTE, 1986
7033	MNW33P	Leyland Fleetline FE30GR	Roe	H43/33F	1976	Ex WYPTE, 1986
7038	MNW38P	Leyland Fleetline FE30GR	Roe	H43/33F	1976	Ex WYPTE, 1986
7043	RWU43R	Leyland Fleetline FE30AGR	Roe	H43/33F	1976	Ex WYPTE, 1986
7046	RWU46R	Leyland Fleetline FE30AGR	Roe	H43/33F	1976	Ex WYPTE, 1986
7054	RWU54R	Leyland Fleetline FE30AGR	Roe	H43/33F	1976	Ex WYPTE, 1986

7073-7088

		Leyland Fleetline FE30AGR	Northern Counties		H43/31F	1979-80 Ex WYPTE, 1986

7073	JNW73V	7079	JUM79V	7082	JUM82V	7086	JUM86V	7088	JUM88V
7078	JUM78V								

7092-7156

		Leyland Fleetline FE30AGR	Roe		H43/33F	1977-78 Ex WYPTE, 1986

7092	WUM92S	7112	WUM112S	7126	WUM126S	7148	CWU148T	7153	CWU153T
7093	WUM93S	7120	WUM120S	7137	CWU137T	7149	CWU149T	7155	CWU155T
7097	WUM97S	7121	WUM121S	7144	CWU144T	7152	CWU152T	7156	CWU156T
7109	WUM109S	7125	WUM125S						

7203-7211

		Leyland Fleetline FE30AGR	Northern Counties		H43/32F	1977-78 Ex Greater Manchester, 1987

7203	PTD640S	7205	PTD646S	7207	PTD649S	7209	PTD651S	7211	PTD658S

7214-7234

		Leyland Fleetline FE30GR	Northern Counties		H43/32F	1978-79 Ex Greater Manchester, 1987

7214	XBU5S	7221	XBU17S	7227	ANA33T	7230	ANA44T	7233	BVR53T
7216	XBU9S	7223	ANA25T	7228	ANA34T	7231	ANA45T	7234	BVR55T
7219	XBU15S	7226	ANA31T	7229	ANA40T	7232	ANA48T		

7235-7243

		Leyland Fleetline FE30AGR	Northern Counties		H43/32F	1978-80 Ex Greater Manchester, 1987

7235	TWH690T	7237	TWH692T	7239	TWH695T	7242	BCB611V	7243	BCB612V
7236	TWH691T	7238	TWH693T	7241	BCB610V				

7244-7253

		Leyland Fleetline FE30GR	Northern Counties		H43/32F	1979 Ex GM Buses, 1988

7244	BVR52T	7246	BVR67T	7248	BVR70T	7250	BVR85T	7252	BVR92T
7245	BVR65T	7247	BVR69T	7249	BVR71T	7251	BVR87T	7253	BVR97T

7511	UWW511X	MCW Metrobus DR101/15	Alexander RH	H43/32F	1982	Ex WYPTE, 1986
7520	UWW520X	MCW Metrobus DR101/15	Alexander RH	H43/32F	1982	Ex WYPTE, 1986

Possibly one of the oldest buses to acquire new branding, Roe-bodied Leyland Atlantean 6030, GUG557N entered service almost 21 -years ago on 21st November 1974, carrying the Metro Leeds insignia. It is seen recently at the interchange with Bradford Traveller fleetname.
Andrew Jarosz

A considerable number of MCW Metrobuses continue in Leeds and are being re-branded with Leeds City Link fleetnames. No.7549, A549KUM, dates from 1984 and is seen in Church Lane, Pudsey showing the first version of the fleetname which is now being superceded by a second version with additional stripes. *Andrew Jarosz*

7521-7537   MCW Metrobus DR102/32   MCW   H46/30F   1983   Ex WYPTE, 1986

7521	CUB521Y	7525	CUB525Y	7531	CUB531Y	7534	CUB534Y	7536	CUB536Y
7522	CUB522Y	7527	CUB527Y	7532	CUB532Y	7535	CUB535Y	7537	CUB537Y
7523	CUB523Y	7530	CUB530Y						

7541-7580   MCW Metrobus DR102/38   MCW   H46/30F   1984   Ex WYPTE, 1986

7541	A541KUM	7550	A750LWY	7558	A758LWY	7566	B566RWY	7572	B572RWY
7542	A542KUM	7551	A751LWY	7560	A760LWY	7567	B567RWY	7575	B575RWY
7544	A544KUM	7552	A752LWY	7561	B561RWY	7568	B568RWY	7577	B577RWY
7545	A545KUM	7553	A753LWY	7562	B562RWY	7569	B569RWY	7578	B578RWY
7546	A546KUM	7554	A754LWY	7563	B563RWY	7570	B570RWY	7579	B579RWY
7547	A547KUM	7555	A755LWY	7564	B564RWY	7571	B571RWY	7580	B580RWY
7549	A549KUM	7556	A756LWY	7565	B565RWY				

7581-7595   MCW Metrobus DR102/66   MCW   H46/31F   1988

7581	F581XWY	7584	F584XWY	7587	F587XWY	7590	F590XWY	7593	F593XWY
7582	F582XWY	7585	F585XWY	7588	F588XWY	7591	F591XWY	7594	F594XWY
7583	F583XWY	7586	F586XWY	7589	F589XWY	7592	F592XWY	7595	F595XWY

7596-7600   MCW Metrobus DR102/67   MCW   H46/31F   1988

7596	F596XWY	7597	F597XWY	7598	F598XWY	7599	F599XWY	7600	F600XWY

7601-7605   MCW Metrobus DR102/69   MCW   H46/31F   1988

7601	F601XWY	7602	F602XWY	7603	F603XWY	7604	F604XWY	7605	F605XWY

8001-8005

8001-8005		Scania N113DRB		Alexander RH		H47/33F		1990	
8001	G801JYG	**8002**	G802JYG	**8003**	G803JYG	**8004**	G804JYG	**8005**	G805JYG

8006-8010

Scania N113DRB		Northern Counties Palatine		H43/33F		1990			
8006	H806TWX	**8007**	H807TWX	**8008**	H808TWX	**8009**	H809TWX	**8010**	H810TWX

8011-8042

Scania N113DRB   Alexander RH   H47/31F   1991

8011	H611VNW	**8018**	H618VNW	**8025**	H625VNW	**8031**	H631VNW	**8037**	H637VNW
8012	H612VNW	**8019**	H619VNW	**8026**	H726VNW	**8032**	H632VNW	**8038**	H638VNW
8013	H613VNW	**8020**	H620VNW	**8027**	H627VNW	**8033**	H633VNW	**8039**	H639VNW
8014	H614VNW	**8021**	H621VNW	**8028**	H628VNW	**8034**	H634VNW	**8040**	H640VNW
8015	H615VNW	**8022**	H622VNW	**8029**	H629VNW	**8035**	H643VNW	**8041**	H641VNW
8016	H616VNW	**8023**	H623VNW	**8030**	H630VNW	**8036**	H636VNW	**8042**	H642VNW
8017	H617VNW	**8024**	H624VNW						

8401-8405

Scania L113CRL   Alexander Strider   B48F   1994

8401	M401UUB	**8402**	M402UUB	**8403**	M403UUB	**8404**	M404UUB	**8405**	M405UUB

8406-8425

Scania L113CRL   Wright Axcess   B48F   1995

8406	N406ENW	**8410**	N410ENW	**8414**	N414ENW	**8418**	N418ENW	**8422**	N422
8407	N407ENW	**8411**	N411ENW	**8415**	N415ENW	**8419**	N419ENW	**8423**	N423
8408	N408ENW	**8412**	N412ENW	**8416**	N416ENW	**8420**	N420	**8424**	N424
8409	N409ENW	**8413**	N413ENW	**8417**	N417ENW	**8421**	N421	**8425**	N425

8534	RWT534R	Leyland Leopard PSU4D/4R		Plaxton Derwent	DP43F	1976	Ex WYPTE, 1986
8600	D727GDE	Scania K92CRB		East Lancashire	B59F	1987	Ex Rhodeservices, 1994

8601-8634

Scania N113CRB   Alexander Strider   B50F   1993

8601	K601HUG	**8608**	K608HUG	**8615**	K615HUG	**8622**	K622HUG	**8629**	K629HUG
8602	K602HUG	**8609**	K609HUG	**8616**	K616HUG	**8623**	K623HUG	**8630**	K630HUG
8603	K603HUG	**8610**	K610HUG	**8617**	K617HUG	**8624**	K624HUG	**8631**	K631HUG
8604	K604HUG	**8611**	K611HUG	**8618**	K618HUG	**8625**	K625HUG	**8632**	K632HUG
8605	K605HUG	**8612**	K612HUG	**8619**	K619HUG	**8626**	K626HUG	**8633**	K633HUG
8606	K606HUG	**8613**	K613HUG	**8620**	K620HUG	**8627**	K627HUG	**8634**	K634HUG
8607	K607HUG	**8614**	K614HUG	**8621**	K621HUG	**8628**	K628HUG		

8635	K1YRL	Scania N113CRB		Alexander Strider	DP50F	1993	

8636-8655

Scania N113CRB   Alexander Strider   B48F   1994

8636	L636PWR	**8640**	L640PWR	**8644**	L644PWR	**8648**	L648PWR	**8652**	L652PWR
8637	L637PWR	**8641**	L641PWR	**8645**	L645PWR	**8649**	L649PWR	**8653**	L653PWR
8638	L638PWR	**8642**	L642PWR	**8646**	L646PWR	**8650**	L650PWR	**8654**	L654PWR
8639	L639PWR	**8643**	L643PWR	**8647**	L647PWR	**8651**	L651PWR	**8655**	L655PWR

Operating companies: Rider York: 228-36/8/90/1, 332/3/5/8/40/63/8/71, 727/34/45/6/52/5, 977/80/8/94, 1171; 1201-8, 1333/49-57/70, 1461, 1547/53/81-97, 1623/56-8, 5186-89/93-7, 5521, 8401-5. Quickstep: 3-72, 1366, 2260-3, 6279/80 At the time of going to press no official allocation of vehicles to the new operating division was available.

Liveries: Cream, green and red (old Yorkshire Rider scheme); cream, gold and red (Gold Rider):1402-7/18/22, 1604/5/13-22 white and red (Quickstep):3-72; blue, silver and red (Superbus):8635-55; black (Jetliner): 1400; National Express: 1415/6/9-21/3-5; traditional liveries: 363, 5196 (York); 1608, 5096 (Halifax); 5156 (Todmorden); 5504/9 (West Yorkshire RCC); 5117/34(Bradford); 6299 (Huddersfield); 7575 (Leeds); 8534 (Todmorden). New liveries are expected to be announced shortly.

Previous Registrations:

23PTA	F418EWR	HUA606Y	A608XYG, WSV410	NIB4908	D404LUA
8995WY	G73RGG	K1YRL	From new	UOI4323	BKH129X
CSU244	REL402R	LOI6690	REL400R	WSV408	HUA604Y
GSU388	E405RWR	NIB4905	D403LUA	WSV409	HUA605Y
GSU390	E407UWR	NIB4906	C402EWU	YR3939	From new

One of the few examples of a Contravision advertisement is shown on Scania 8004, G804JYG, here in the colours of Barclays Mastercard. Little light is lost inside despite its apparent opaqueness when viewed from outside.
Phillip Stephenson

Below: **Prior to the opening of the Leeds guided busway, for which they were purchased, Alexander Strider-bodied Scania 8651, L651PWR was seen outside the main Post Office en route for Pudsey.**
Andrew Jarosz

Newly delivered and with SMT Lothians names are ten Mercedes-Benz O405s with Optare Prisma bodies. Show on a test run is 66, N66CSC which carries the Gold Service livery. To accommodate the batch several of the Leyland Fleetlines have been renumbered in the 800 series. *Andrew Jarosz*

Princes Street, Edinburgh is the location of this picture of 337, WSV137. This Leyland Tiger with Plaxton Paramount bodywork carries the SMT Coaches name on fleet livery that has replaced a Scottish Citylink scheme. *Tony Wilson*

SMT

Eastern Scottish Omnibuses Ltd, 2 Westfield Avenue, Edinburgh EH11 2QH

Depots : Eskbank Road, Dalkeith; Westfield Avenue, Edinburgh; Almondvale, Livingston; Deans, Livingston and The Mall, Musselburgh.

2	J812KHD	DAF SB3000DKV601	Van Hool Alizée	C49FT	1992	Ex ?, 19
5	J795KHD	DAF MB230LB615	Plaxton Paramount 3500 III	C51FT	1992	Ex ?, 19
8	J8SMT	Dennis Javelin 12SDA1929	Plaxton Paramount 3200 III	C53F	1992	
9	L109OSX	Dennis Javelin 12SDA2131	Plaxton Premiére 320	C53F	1994	
10	L110OSX	Dennis Javelin 12SDA2131	Plaxton Premiére 320	C53F	1994	
19	A19SMT	Dennis Javelin 12SDA1907	Duple 320	C53FT	1988	Ex Martin, Spean Bridge, 1992
23	AWG623	AEC Regal	Alexander	C31F	1947	Ex preservation, 1986

33-41

		Ailsa B55-10		Alexander AV		H44/35F	1978	Ex Clydeside Scottish, 1988

33	TSJ593S	36	TSJ596S	38	TSJ598S	40	TSJ600S	41	TSJ601S
34w	TSJ594S	37	TSJ597S	39	TSJ599S				

42-50

		Ailsa B55-10		Alexander AV		H44/35F	1978	Ex Central, 1988

42	BGG252S	44	BGG254S	46	BGG256S	48w	BGG258S	50	BGG260S
43w	BGG253S	45w	BGG255S	47	BGG257S	49w	BGG259S		

61-70

		Mercedes-Benz O405		Optare Prisma		B49F	1995

61	N61CSC	63	N63CSC	65	N65CSC	67	N67CSC	69	N69CSC
62	N62CSC	64	N64CSC	66	N66CSC	68	N68CSC	70	N70CSC

76-95

		Volvo B55-10 MkIII		Alexander RV		H44/35F	1981

76	HSF76X	81	HSF81X	85	HSF85X	88	HSF88X	93	HSF93X
77	HSF77X	82	HSF82X	86	HSF86X	91	HSF91X	94	HSF94X
78	HSF78X	83	HSF83X	87	HSF87X	92	HSF92X	95	HSF95X
80	HSF80X	84	HSF84X						

100-115

		Leyland Olympian ONLXB/1R		Eastern Coach Works		H45/32F*	1982	*101-4 are DPH45/32F

100	ULS100X	104	ULS104X	107	ULS107X	110	ULS110X	113	ULS113X
101	ULS101X	105	ULS105X	108	ULS108X	111	ULS111X	114	ULS114X
102	ULS102X	106	ULS106X	109	ULS109X	112	ULS112X	115	ULS115X
103	ULS103X								

117-142

		Leyland Olympian ONLXB/1R		Alexander RL		H45/32F	1983-84

117	ALS117Y	124	ALS124Y	129	ALS129Y	135	ALS135Y	139	A139BSC
118	ALS118Y	125	ALS125Y	132	ALS132Y	136	A136BSC	140	A140BSC
119	ALS119Y	126	ALS126Y	133	ALS133Y	137	A137BSC	141	A141BSC
122	ALS122Y	127	ALS127Y	134	ALS134Y	138	A138BSC	142	A142BSC
123	ALS123Y	128	ALS128Y						

149-158

		Volvo B55-10 MkIII		Alexander RV		H44/37F	1984

149	B149GSC	151	B151GSC	153	B153GSC	155	B155GSC	157	B157GSC
150	B150GSC	152	B152GSC	154	B154GSC	156	B156GSC	158	B158GSC

161	B161KSC	Leyland Olympian ONTL11/1R	Alexander RL	H45/32F	1985
162	B162KSC	Leyland Olympian ONTL11/1R	Alexander RL	H45/32F	1985
163	B163KSC	Leyland Olympian ONTL11/1R	Alexander RL	H45/32F	1985

169-173
Volvo Citybus B10M-50 — Alexander RV — H44/37F — 1985

169	B169KSC	170	B170KSC	171	B171KSC	172	B172KSC	173	B173KSC

174-186
Leyland Lion LDTL11/2R — Alexander RH — DPH49/37F* — 1986-7 — *180/4-6 are DPH45/35F

174	C174VSF	177	C177VSF	180	C180VSF	183	C183VSF	185	D185ESC
175	C175VSF	178	C178VSF	181	C181VSF	184	D184ESC	186	D186ESC
176	C176VSF	179	C179VSF	182	C182VSF				

187	E187HSF	Volvo Citybus B10M-50	Alexander RV	DPH45/35F	1987
188	E188HSF	Volvo Citybus B10M-50	Alexander RV	DPH45/35F	1987
189	E189HSF	Volvo Citybus B10M-50	Alexander RV	DPH45/35F	1987
190	E190HSF	Volvo Citybus B10M-50	Alexander RV	DPH45/35F	1987

201-212
Volvo B10B-58 — Alexander Strider — B51F — 1993

201	L201KFS	204	L204KSX	207	L207KSX	209	L209KSX	211	L211KSX
202	L202KFS	205	L205KSX	208	L208KSX	210	L210KSX	212	L212KSX
203	L203KSX	206	L206KSX						

213	L213KSX	Volvo B10B-58	Wright Endurance	B51F	1993

232-237
Leyland Olympian ONLXB/1R — East Lancashire — H47/32F — 1984 — Ex Grampian, 1994

232	A77RRP	234	A79RRP	235	A80RRP	236	A81RRP	237	A82RRP
233	A78RRP								

311	PSF311Y	Leyland Tiger TRBTL11/2R	Alexander AT	C49F	1982
312	PSF312Y	Leyland Tiger TRBTL11/2R	Alexander AT	C49F	1982
328	A328BSC	Leyland Tiger TRBTL11/2RP	Alexander TE	C49F	1983
329	A329BSC	Leyland Tiger TRBTL11/2RP	Alexander TE	C49F	1983
330	WSV140	Leyland Tiger TRCTL11/2R	Plaxton Paramount 3200 E	C49F	1984
331	A20SMT	Leyland Tiger TRCTL11/2R	Plaxton Paramount 3200 E	C49F	1984
332	WSV135	Leyland Tiger TRCTL11/2RH	Plaxton Paramount 3200 E	C49F	1984

335-341
Leyland Tiger TRCTL11/2RH — Plaxton Paramount 3200 II — C49F — 1985

335	WSV144	337	WSV137	339	A9SMT	340	A10SMT	341	B341RLS
336	WSV136	338	WSV138						

344w	A14SMT	Leyland Tiger TRCTL11/3RZ	Duple Caribbean 2	C51F	1984	Ex Duple demonstrator, 1986
345	A15SMT	Leyland Tiger TRCTL11/3RH	Duple 340	C53F	1987
346	A16SMT	Leyland Tiger TRCTL11/3RH	Duple 340	C53F	1987
347	A17SMT	Leyland Tiger TRCTL11/3RH	Duple 340	C53F	1987
348	A18SMT	Leyland Tiger TRCTL11/3RH	Duple 340	C53F	1987
349	D349ESC	Leyland Tiger TRBTL11/2RP	Alexander TE	C49F	1987
350	D350ESC	Leyland Tiger TRBTL11/2RP	Alexander TE	C49F	1987
351	D351ESC	Leyland Tiger TRBTL11/2RP	Alexander TE	C49F	1987
352	A12SMT	Leyland Tiger TRCL10/3RZA	Duple 340	C53F	1987
353	A13SMT	Leyland Tiger TRCL10/3RZA	Duple 340	C53F	1987
399	M399OMS	Omni	Omni Citizen	B16FL	1994

404-430
Renault-Dodge S56 — Alexander AM — B21F* — 1986 — *420 is B16FL

404w	D404ASF	411w	D411ASF	416w	D416ASF	422w	D421ASF	426w	D426ASF
405w	D405ASF	412w	D412ASF	418w	D418ASF	422w	D422ASF	427w	D427ASF
406w	D406ASF	413w	D413ASF	419w	D419ASF	423w	D423ASF	428w	D428ASF
409w	D409ASF	414w	D414ASF	420	D420ASF	424w	D424ASF	429w	D429ASF

Opposite, top: **'Citylinking' into Edinburgh is 352, A12SMT, a Leyland Tiger with Duple 340 bodywork. This is one of five vehicles currently wearing the Citylink livery. Cunningly, SMT purchased several of the A-prefix index marks with their initals while they were available from DVLA at the standard rate, rather than the inflated prices these numbers may attract.** *Tony Wilson*
Opposite, bottom: **Originally new to Northampton, Leyland Olympian 232, A77RRP arrived with SMT from the Grampian fleet. This vehicle features in that livery on the cover of the current edition of the Scottish Bus Handbook which covers the major, and not so major, operators in Scotland.** *Tony Wilson*

431-470 — Renault-Dodge S56 — Alexander AM — B25F* — 1987 — *466/8-70 are DP25F

431	E431JSG	443	E443JSG	451	E451JSG	457w	E457JSG	463	E463JSG
432	E432JSG	444	E444JSG	452w	E452JSG	458	E458JSG	464w	E464JSG
436	E436JSG	446w	E446JSG	453	E453JSG	459	E459JSG	466	E466JSG
437	E437JSG	447	E447JSG	454	E454JSG	460	E460JSG	468	E468JSG
438	E438JSG	448	E448JSG	455	E455JSG	461	E461JSG	469	E469JSG
439	E439JSG	449	E449JSG	456	E456JSG	462	E462JSG	470	E470JSG
442	E442JSG	450	E450JSG						

473-501 — Renault S75 — Reeve Burgess Beaver — B31F — 1991

473	H473OSC	485	H485OSC	489	H489OSC	493	H493OSC	497	H497OSC
474	H474OSC	486	H486OSC	490	H490OSC	494	H494OSC	498	H498OSC
477	H477OSC	487	H487OSC	491	H491OSC	495	H495OSC	499	H499OSC
478	H478OSC	488	H488OSC	492	H492OSC	496	H496OSC	501	H501OSC
479	H479OSC								

503-517 — Optare MetroRider — Optare — B31F — 1992

503	J503WSX	506	J506WSX	509	J509WSX	512	J512WSX	515	K515BSX
504	J504WSX	507	J507WSX	510	J510WSX	513	K513BSX	516	K516BSX
505	J505WSX	508	J508WSX	511	J511WSX	514	K514BSX	517	K517BSX

518-527 — Optare MetroRider — Optare — B31F — 1994

518	L518KSX	520	L520KSX	522	L522KSX	525	L525KSX	527	L527KSX
519	L519KSX	521	L521KSX	524	L524KSX	526	L526KSX		

568w	B568LSC	Leyland Tiger TRCTL11/3RH	Duple Caribbean 2	C49F	1985
569w	B569LSC	Leyland Tiger TRCTL11/3RH	Duple Caribbean 2	C49F	1985

591-660 — Seddon Pennine 7 — Alexander AYS — B53F* — 1979-80 — *608/60 are DP49F; 634 is B47F
604/14/33/4 ex Midland Bluebird, 1994; 608/9/66 ex Lowland, 1995

591	RSX591V	612	SSX612V	623	SSX623V	636	YSG636W	647	YSG647W
593w	RSX593V	613	SSX613V	624	SSX624V	637	YSG637W	650	YSG650W
594	RSX594V	614w	SSX614V	625	SSX625V	638	YSG638W	654	YSG654W
595	SSX595V	616	SSX616V	627	SSX627V	640	YSG640W	655	YSG655W
596	SSX596V	617	SSX617V	628	SSX628V	641	YSG641W	656	YSG656W
597w	SSX597V	618	SSX618V	629	SSX629V	642	YSG642W	657	YSG657W
604	SSX604V	619	SSX619V	630	SSX630V	643	YSG643W	658	YSG658W
608	SSX608V	620	SSX620V	633	YSG633W	644	YSG644W	659	YSG659W
609	SSX609V	621	SSX621V	634	YSG634W	645	YSG645W	660	YSG660W
611w	SSX611V	622	SSX622V	635	YSG635W	646	YSG646W		

Two-dozen Optare MetroRiders operate with SMT. Seen heading for Tranent is 521, L521KSX, which carries SMT Lothian names. Typical of the former GRT Bus Group names, a thistle motif is used as the dot over the i, although this, like Badgerline Group's badger, is being deleted in FirstBus guise.
Richard Walter

701	ORS201R	Leyland Atlantean AN68A/1R	Alexander AL	H45/34F	1976	Ex Midland Bluebird, 1994		
712	ORS212R	Leyland Atlantean AN68A/1R	Alexander AL	H45/34F	1976	Ex Midland Bluebird, 1995		
726	XSA226S	Leyland Atlantean AN68A/1R	Alexander AL	H45/34F	1978	Ex Midland Bluebird, 1994		
733	YSO233T	Leyland Altantean AN68A/1R	Alexander AL	H45/34F	1979	Ex Grampian, 1995		

773-782

Ailsa B55-10 — Alexander AV — H44/35F — 1978

773	CSG773S	775	CSG775S	777	CSG777S	779W	CSG779S	782w	CSG782S
774	CSG774S	776	CSG776S	778	CSG778S	780	CSG780S		

792	KSA192P	Leyland Atlantean AN68A/1R	Alexander AL	H45/34F	1976	Ex Midland Bluebird, 1994

852-875

Leyland Fleetline FE30AGR — Eastern Coach Works — H43/32F — 1978-79

852w	GSC852T	859	GSC859T	862	OSG62V	869	OSG69V	871	OSG61V
854	OSG54V	860w	GSC860T	865	OSG65V	870	OSG60V	875	OSG75V
855w	GSC855T	861	GSC861T	866	OSG66V				

896-922

Seddon Pennine 7 — Alexander AT — C49F — 1978-79

896	GSX896T	898	GSX898T	900	GSX900T	901w	GSX901T	922w	JSF922T
897	GSX897T							

932	LSC932T	Seddon Pennine 7	Alexander AYS	B53F	1979	
933	LSC933T	Seddon Pennine 7	Alexander AYS	B53F	1979	
938	LSC938T	Seddon Pennine 7	Alexander AYS	B53F	1979	
939w	LSC939T	Seddon Pennine 7	Alexander AYS	B53F	1979	
979	JFS979X	Seddon Pennine 7	Alexander AYS	B53F	1982	
980	JFS980X	Seddon Pennine 7	Alexander AYS	B53F	1982	
981	JFS981X	Seddon Pennine 7	Alexander AYS	B53F	1982	
982	JFS982X	Seddon Pennine 7	Alexander AYS	B53F	1982	
983	JFS983X	Seddon Pennine 7	Alexander AYS	B49F	1982	Ex Midland Bluebird, 1994

Previous Registrations:

A9SMT	B339RLS	A17SMT	D347ESC	WSV135	A332BSC
A10SMT	B340RLS	A18SMT	D348ESC	WSV136	B336RLS
A12SMT	E352KSF	A19SMT	F250OFP	WSV137	B337RLS
A13SMT	E353KSF	A20SMT	A331BSC	WSV138	B338RLS
A14SMT	B466WRN	AWG623	From new	WSV140	A330BSC
A15SMT	D345ESC	J8SMT	J864WSC	WSV144	B335RLS, GCS245
A16SMT	D346ESC				

Livery: Cream and green; red and yellow (coaches) 8/19, 337/8/40/4; yellow and blue (Scottish Citylink) 9/10, 345/6/52.

The double deck buses of SMT are represented by 169, B169KSC, one of five Volvo B10M Citybuses for normal service work delivered in 1985 - the further four examples from 1987 feature high-back seating. Photographed in St Andrews Square, Edinburgh, it is seen heading for Loanhead.
Tony Wilson

SWT

South Wales Transport Co Ltd, Heol Gwyrosydd, Penlan, Swansea, SA5 7BN

Depots: Withybush Industrial Estate, Haverfordwest; Inkerman Street, Llanelli; Tawe Terrace, Pontardawe and Pentregethin Road, Ravenhill, Swansea. **Outstations:** Ammanford, Carmarthen, Gorseinon, Milford Haven, Neath, Pembroke Dock and Tenby

101	J1SWT	Volvo B10M-60	Plaxton Expressliner	C46FT	1991
102	J2SWT	Volvo B10M-60	Plaxton Expressliner 2	C46FT	1992
103	J3SWT	Volvo B10M-60	Plaxton Expressliner 2	C46FT	1992
104	J4SWT	Volvo B10M-60	Plaxton Expressliner 2	C46FT	1992
105	J5SWT	Volvo B10M-60	Plaxton Expressliner 2	C46FT	1992
106	L506GEP	Volvo B10M-60	Plaxton Expressliner 2	C46FT	1993
107	M107NEP	Dennis Javelin GX 12SDA2132	Plaxton Expressliner 2	C44FT	1994
108	M108NEP	Dennis Javelin GX 12SDA2132	Plaxton Expressliner 2	C44FT	1994
109	M109PWN	Dennis Javelin GX 12SDA2133	Plaxton Expressliner 2	C44FT	1995
110	M110PWN	Dennis Javelin GX 12SDA2133	Plaxton Expressliner 2	C44FT	1995
111	M111PWN	Dennis Javelin GX 12SDA2133	Plaxton Expressliner 2	C44FT	1995
131	C312KTH	Hestair-Duple SDA1510	Duple 425	C53F	1986
132	ACY178A	Hestair-Duple SDA1510	Duple 425	C48FT	1986
133	MKH487A	Hestair-Duple SDA1510	Duple 425	C48FT	1986
134	F134DEP	Hestair-Duple SDA1510	Duple 425	C48FT	1989
135	F135DEP	Hestair-Duple SDA1510	Duple 425	C48FT	1989
136	F99CEP	Hestair-Duple SDA1510	Duple 425	C46FT	1989	Ex United Welsh Coaches, 1989
137	F100CEP	Hestair-Duple SDA1512	Duple 425	C46FT	1989	Ex Brewers, 1989
139	E206BOD	Hestair-Duple SDA1510	Duple 425	C50FT	1988	Ex Brewers, 1995

215-253

	Mercedes-Benz L608D	Robin Hood	B20F*	1985-86 *220 is DP19F

215	C215HTH	226	D226LCY	234	D234LCY	241	D241LCY	248	D248LCY
220	D220LCY	227	D227LCY	235	D235LCY	242	D242LCY	249	D249LCY
221	D221LCY	228	D228LCY	236	D236LCY	243	D243LCY	250	D250LCY
222	D222LCY	229	D229LCY	237	D237LCY	244	D244LCY	251	D251LCY
223	D223LCY	230	D230LCY	238	D238LCY	245	D245LCY	252	D252LCY
224	D224LCY	232	D232LCY	239	D239LCY	246	D246LCY	253	D253LCY
225	D225LCY	233	D233LCY	240	D240LCY	247	D247LCY		

279	E279TTH	Mercedes-Benz 609D	Robin Hood	B20F	1987
280	E280TTH	Mercedes-Benz 609D	Robin Hood	B20F	1987
281	E281TTH	Mercedes-Benz 609D	Robin Hood	B20F	1987
282	E282TTH	Mercedes-Benz 609D	Robin Hood	B20F	1987

287-322

	Mercedes-Benz 709D	Reeve Burgess Beaver	B25F*	1988	287/9 ex Brewers, 1994; *304 is DP25F

287	E287UCY	295	E295VEP	302	E302VEP	309	F309AWN	316	F316AWN
289	E289VEP	296	E296VEP	303	E303VEP	310	F310AWN	317	F317AWN
290	E290VEP	297	E297VEP	304	E304VEP	311	F311AWN	318	F318AWN
291	E291VEP	298	E298VEP	305	E305VEP	312	F312AWN	319	F319AWN
292	E292VEP	299	E299VEP	306	E306VEP	313	F313AWN	320	F320AWN
293	E293VEP	300	E300VEP	307	F307AWN	314	F314AWN	321	F321AWN
294	E294VEP	301	E301VEP	308	F308AWN	315	F315AWN	322	F322AWN

323-327

	Mercedes-Benz 811D	Reeve Burgess Beaver	B25F	1989	Ex Brewers, 1994

323	F323DCY	324	F324DCY	325	F325DCY	326	F326DCY	327	F327DCY

328-347

	Mercedes-Benz 814D	Robin Hood	B31F	1989

328	F328FCY	332	F332FCY	336	F336FCY	340	F340FCY	344	G344GEP
329	F329FCY	333	F333FCY	337	F337FCY	341	F341FCY	345	G345GEP
330	F330FCY	334	F334FCY	338	F338FCY	342	F342FCY	346	G346GEP
331	F331FCY	335	F335FCY	339	F339FCY	343	F343FCY	347	G347GEP

348-361

	Mercedes-Benz 814D	Phoenix	B31F	1989

348	G348JTH	351	G351JTH	354	G354JTH	357	G357JTH	360	G360JTH
349	G349JTH	352	G352JTH	355	G355JTH	358	G358JTH	361	G361JTH
350	G350JTH	353	G353JTH	356	G356JTH	359	G359JTH		

The South Wales to Heathrow and Gatwick National Express service is now marketed as one of the AirLink services in a silver, blue, black and red livery. Normally used on services from the major airports and seen here when new, SWT 111, M111PWN, is heading for London Victoria. *Cliff Beeton*

Phoenix bodywork is fitted to the 1989 and 1990 deliveries of Mercedes-Benz 814s. Shown with Valley Link names is 360, G360JTH seen as it arrives in Swansea on service X24. *John Jones*

365-371

Mercedes-Benz 814D   Phoenix   B31F*   1990   *367-70 are DP31F

365	G365JTH	367	G367MEP	369	G369MEP	370	G370MEP	371	G371MEP
366	G366JTH	368	G368MEP						

372	G372MEP	Mercedes-Benz 814D	Plaxton Beaver (1992)	B31F	1990

373-381

Mercedes-Benz 814D   Phoenix   B31F   1990

373	G373MEP	375	H375OTH	377	H377OTH	379	H379OTH	381	H381OTH
374	H374OTH	376	H376OTH	378	H378OTH	380	H380OTH		

382	H382TTH	Mercedes-Benz 814D	Reeve Burgess Beaver	B31F	1991	
500	WNO484	Bristol KSW5G	Eastern Coach Works	O33/28R	1953	Ex Eastern National, 1974

501-524

Dennis Dart 9SDL3034 — Plaxton Pointer — B31F — 1993

501	L501HCY	506	L506HCY	511	L511HCY	516	L516HCY	521	L521HCY
502	L502HCY	507	L507HCY	512	L512HCY	517	L517HCY	522	L522HCY
503	L503HCY	508	L508HCY	513	L513HCY	518	L518HCY	523	L523HCY
504	L504HCY	509	L509HCY	514	L514HCY	519	L519HCY	524	L524HCY
505	L505HCY	510	L510HCY	515	L515HCY	520	L520HCY		

525-550

Dennis Dart 9SDL3034 — Plaxton Pointer — B31F — 1994

525	L525JEP	531	L531JEP	536	L536JEP	541	L541JEP	546	L546JEP
526	L526JEP	532	L532JEP	537	L537JEP	542	L542JEP	547	L547JEP
527	L527JEP	533	L533JEP	538	L538JEP	543	L543JEP	548	L548JEP
528	L528JEP	534	L534JEP	539	L539JEP	544	L544JEP	549	L549JEP
529	L529JEP	535	L535JEP	540	L540JEP	545	L545JEP	550	L550JEP
530	L530JEP								

551-568

Dennis Dart 9SDL3034 — Plaxton Pointer — B31F — 1995

551	N551UCY	555	N555UCY	559	N559UCY	563	N563UCY	566	N566UCY
552	N552UCY	556	N556UCY	561	N561UCY	564	N564UCY	567	N567UCY
553	N553UCY	557	N557UCY	562	N562UCY	565	N565UCY	568	N568UCY
554	N554UCY	558	N558UCY						

607	F607AWN	Mercedes-Benz 709D	Reeve Burgess Beaver	DP25F	1988	
791	OEP791R	Leyland National 11351A/1R		B49F	1976	Ex Brewers, 1995
792	OEP792R	Leyland National 11351A/1R		B49F	1976	Ex Brewers, 1995
802	TWN802S	Leyland National 11351A/1R		B49F	1978	Ex Brewers, 1995
804	WWN804T	Leyland National 11351A/1R		B49F	1979	Ex Brewers, 1995
805	WWN805T	Leyland National 11351A/1R		B49F	1979	Ex Brewers, 1995
813	AWN813V	Leyland National 11351A/1R		B49F	1979	
815	AWN815V	Leyland National 11351A/1R		B52F	1979	

816-825

Dennis Lance 11SDA3112 — Plaxton Verde — DP45F — 1993

816	L816HCY	818	L818HCY	820	L820HCY	822	L822HCY	824	L824HCY
817	L817HCY	819	L819HCY	821	L821HCY	823	L823HCY	825	L825HCY

901-907

Leyland Olympian ONCL10/1RV — Eastern Coach Works — H45/30F — 1985

901	C901FCY	903	C903FCY	905	C905FCY	906	C906FCY	907	C907FCY
902	C902FCY	904	C904FCY						

908	IIL1828	Leyland Olympian ONTL11/2RSp Eastern Coach Works	CH45/28F	1985	Ex Thamesway, 1991
909	IIL1829	Leyland Olympian ONTL11/2RSp Eastern Coach Works	CH45/28F	1985	Ex Thamesway, 1991

975-995

Bristol VRT/SL3/501 — Eastern Coach Works — H43/31F — 1979-80

975	BEP975V	980	BEP980V	989	ECY989V	991	ECY991V	994	EWN994W
976	BEP976V	984	BEP984V	990	ECY990V	992	EWN992W	995	EWN995W
978	BEP978V								

7500-7569

Mercedes-Benz L608D — Dormobile — B20F — 1986 — Ex City Line, 1994-95

7500	D500FAE	7542	D542FAE	7545	D545FAE	7552	D552FAE	7559	D559FAE
7515	D515FAE	7543	D543FAE	7546	D546FAE	7557	D557FAE	7565	D565FAE
7533	D533FAE	7544	D544FAE	7547	D547FAE	7558	D558FAE	7569	D569FAE

Named vehicles: 500 *William Gammon*; 908 *Dylan Thomas*; 909 *Sir Harry Secombe*

Previous Registrations:

ACY178A	from new	IIL1828	B690BPU	MKH487A	from new
C312KTH	999BCY	IIL1829	B696BPU	MKH98A	B130CTH

Livery: Green, red, yellow & white; white (National Express) 101-6, 131-3/5-7; silver, blue, black & red (Air Link) 107-11.

Opposite: **No M-registration buses were added to the SWT fleet although five Expressliners were added to the coach fleet. The last deliveries of buses prior to this autumns 17 Darts were similar vehicles in 1994 such as** *(top)* **550, L550JEP working one of the Swansea City services which are colour coded on destinations, timetables and bus stops. 1993 intake included Dennis Lances bodied by Plaxton and featuring 45 high-back seats for the then new** *Timecutter* **services to and from Swansea. No.824, L824HCY is seen passing through Neath.** *Cliff Beeton/John Jones*

THAMESWAY

Thamesway Ltd, Office 24, Eastgate Business Centre, Basildon, SS14 1EB

Depots and outstations: Cherrydown, Basildon; North Rd, Brentwood; London Rd, Hadleigh; Morson Road, Ponders End.

33-37

		Mercedes-Benz L608D		Dormobile	B20F	1986	Ex City Line, 1995
33	D504FAE	**34**	D505FAE	**35**	D506FAE	**36**	D507FAE	**37**	D511FAE

201	C201HJN	Mercedes-Benz L608D	Reeve Burgess	B20F	1985	Ex Eastern National, 1990
202	C696ECV	Mercedes-Benz L608D	Reeve Burgess	DP19F	1985	Ex Western National, 1993
203	C963GCV	Mercedes-Benz L608D	Reeve Burgess	DP19F	1986	Ex Western National, 1994
204	C978GCV	Mercedes-Benz L608D	PMT	B20F	1986	Ex Western National, 1994
206	C957GAF	Mercedes-Benz L608D	Reeve Burgess	B20F	1986	Ex Western National, 1994
207	C699ECV	Mercedes-Benz L608D	Reeve Burgess	DP19F	1985	Ex Western National, 1994
208	C491HCV	Mercedes-Benz L608D	Reeve Burgess	DP19F	1986	Ex Western National, 1994
209	C967GCV	Mercedes-Benz L608D	Reeve Burgess	DP19F	1986	Ex Western National, 1994

225-234

		Mercedes-Benz L608D		Reeve Burgess	B20F	1986	Ex Eastern National, 1990
225	D225PPU	**231**	D231PPU	**232**	D232PPU	**233**	D233PPU	**234**	D234PPU
230	D230PPU						

235-244

		Mercedes-Benz L608D		Dormobile	B20F	1986	Ex Eastern National, 1990
235	D235PPU	**237**	D237PPU	**239**	D239PPU	**241**	D241PPU	**243**	D243PPU
236	D236PPU	**238**	D238PPU	**240**	D240PPU	**242**	D242PPU	**244**	D244PPU

245-260

		Mercedes-Benz 709D		Reeve Burgess Beaver	B23F	1988-89	Ex Eastern National, 1990
245	F245MVW	**249**	F249NJN	**252**	F252NJN	**255**	F255RHK	**258**	F258RHK
246	F246MVW	**250**	F250NJN	**253**	F253RHK	**256**	F256RHK	**259**	F259RHK
247	F247NJN	**251**	F251NJN	**254**	F254RHK	**257**	F257RHK	**260**	F260RHK
248	F248NJN						

261	D764KWT	Mercedes-Benz 609D	Robin Hood	B20F	1987	Ex SWT, 1994

301-356

		Mercedes-Benz 709D		Reeve Burgess Beaver	B23F	1991	
301	H301LPU	**313**	H314LJN	**324**	H331LJN	**335**	H344LJN	**346**	H356LJN
302	H302LPU	**314**	H315LJN	**325**	H332LJN	**336**	H345LJN	**347**	H357LJN
303	H303LPU	**315**	H317LJN	**326**	H334LJN	**337**	H346LJN	**348**	H358LJN
304	H304LPU	**316**	H319LJN	**327**	H335LJN	**338**	H347LJN	**349**	H359LJN
305	H305LPU	**317**	H321LJN	**328**	H336LJN	**339**	H348LJN	**350**	H361LJN
306	H306LPU	**318**	H322LJN	**329**	H337LJN	**340**	H349LJN	**351**	H362LJN
307	H307LJN	**319**	H324LJN	**330**	H338LJN	**341**	H351LJN	**352**	H363LJN
308	H308LJN	**320**	H326LJN	**331**	H339LJN	**342**	H352LJN	**353**	H364LJN
309	H310LJN	**321**	H327LJN	**332**	H341LJN	**343**	H353LJN	**354**	H365LJN
310	H311LJN	**322**	H329LJN	**333**	H342LJN	**344**	H354LJN	**355**	H366LJN
311	H312LJN	**323**	H330LJN	**334**	H343LJN	**345**	H355LJN	**356**	H367LJN
312	H313LJN						

357-387

		Mercedes-Benz 709D		Reeve Burgess Beaver	B23F	1991	
357	H368OHK	**364**	H375OHK	**370**	H381OHK	**376**	H387OHK	**382**	H393OHK
358	H369OHK	**365**	H376OHK	**371**	H382OHK	**377**	H388OHK	**383**	H394OHK
359	H370OHK	**366**	H377OHK	**372**	H383OHK	**378**	H389OHK	**384**	H395OHK
360	H371OHK	**367**	H378OHK	**373**	H384OHK	**379**	H390OHK	**385**	H396OHK
361	H372OHK	**368**	H379OHK	**374**	H385OHK	**380**	H391OHK	**386**	H397OHK
362	H373OHK	**369**	H380OHK	**375**	H386OHK	**381**	H392OHK	**387**	H398OHK
363	H374OHK						

388-395

		Mercedes-Benz 709D		Reeve Burgess Beaver	B23F	1991	Ex Eastern National, 1992
388	H388MAR	**390**	H390MAR	**392**	H392MAR	**394**	H394MAR	**395**	H395MAR
389	H389MAR	**391**	H391MAR	**393**	H393MAR			

396	K396KHJ	Mercedes-Benz 709D	Plaxton Beaver	B23F	1993	
397	K397KHJ	Mercedes-Benz 709D	Plaxton Beaver	B23F	1993	

398	K398KHJ	Mercedes-Benz 709D	Plaxton Beaver	B23F	1993	
511	ATH110T	Leyland Leopard PSU5C/4R	Duple 320(1989)	C53F	1979	Ex United Welsh Coaches, 1992
512w	B336BGL	Leyland Tiger TRCTL11/3RH	Duple Caribbean 2	C48FT	1985	Ex Western National, 1992
513w	B337BGL	Leyland Tiger TRCTL11/3RH	Duple Caribbean 2	C46FT	1985	Ex Western National, 1992
514	E675UNE	Leyland Tiger TRCTL11/3RZ	Plaxton Paramount 3200 III	C53F	1988	Ex Shearings, 1992
515	E677UNE	Leyland Tiger TRCTL11/3RZ	Plaxton Paramount 3200 III	C53F	1988	Ex Shearings, 1992
516	F771GNA	Leyland Tiger TRCTL11/3ARZA	Plaxton Paramount 3200 III	C53F	1989	Ex Shearings, 1993
517	J45SNY	Leyland Tiger TRCL10/3ARZM	Plaxton 321	C53F	1991	Ex Bebb, Llantwit Fardre, 1992
518	J46SNY	Leyland Tiger TRCL10/3ARZM	Plaxton 321	C53F	1991	Ex Bebb, Llantwit Fardre, 1992
519	J48SNY	Leyland Tiger TRCL10/3ARZM	Plaxton 321	C53F	1991	Ex Bebb, Llantwit Fardre, 1992
520	J54SNY	Leyland Tiger TRCL10/3ARZM	Plaxton 321	C53F	1991	Ex Bebb, Llantwit Fardre, 1992
521	F613XWY	Leyland Tiger TRCTL11/3R	Plaxton Paramount 3200 E	C53F	1988	Ex Yorkshire Rider, 1994
522	F614XWY	Leyland Tiger TRCTL11/3R	Plaxton Paramount 3200 E	C53F	1988	Ex Yorkshire Rider, 1994

601-619

Volvo B10M-62 — Plaxton Premiére 320 — C53F — 1995-96

601	N601APU	605	N605APU	609	N609APU	613	N613APU	617	N617APU
602	N602APU	606	N606APU	610	N610APU	614	N614APU	618	N618APU
603	N603APU	607	N607APU	611	N611APU	615	N615APU	619	N619APU
604	N604APU	608	N608APU	612	N612APU	616	N616APU		

701	N701CPU	Dennis Dart SLF 10m	Plaxton Pointer	B40F	1995	
705	C483BFB	Ford Transit 190	Dormobile	B16F	1985	Ex Badgerline, 1991

800-804

Mercedes-Benz 811D — Reeve Burgess Beaver — B23F — 1989 — Ex Eastern National, 1990

800	F800RHK	801	F801RHK	802	F802RHK	803	F803RHK	804	F804RHK

805-811

Mercedes-Benz 811D — Plaxton Beaver — B31F — 1992

805	K805DJN	807	K807DJN	809	K809DJN	810	K810DJN	811	K811DJN
806	K806DJN	808	K808DJN						

851	N851CPU	Dennis Dart 9SDL3040	Marshall C36	DP35FL	1995
852	N852CPU	Dennis Dart 9SDL3040	Marshall C36	DP35FL	1995
853	N853CPU	Dennis Dart 9SDL3040	Marshall C36	DP35FL	1995
854	N854CPU	Dennis Dart 9SDL3040	Marshall C36	DP35FL	1995

901-917

Dennis Dart 9SDL3016 — Plaxton Pointer — B35F — 1992

901	K901CVW	905	K905CVW	909	K909CVW	912	K912CVW	915	K915CVW
902	K902CVW	906	K906CVW	910	K910CVW	913	K913CVW	916	K916CVW
903	K903CVW	907	K907CVW	911	K911CVW	914	K914CVW	917	K917CVW
904	K904CVW	908	K908CVW						

Thamesway introduced a purple and yellow livery during 1994, the first to be delivered being a batch of Dennis Darts including 936, M936TEV. As we go to press it has just been announced that Thamesway has been successful in securing LRT services 191 and 399 for which further Dennis Darts will be required.
Keith Grimes

918-943

Dennis Dart 9.8SDL3035   Plaxton Pointer   B39F   1994

918	M918TEV	924	M924TEV	930	M930TEV	935	M935TEV	940	M940TEV
919	M919TEV	925	M925TEV	931	M931TEV	936	M936TEV	941	M941TEV
920	M920TEV	926	M926TEV	932	M932TEV	937	M937TEV	942	M942TEV
921	M921TEV	927	M927TEV	933	M933TEV	938	M938TEV	943	M943TEV
922	M922TEV	928	M928TEV	934	M934TEV	939	M939TEV		
923	M923TEV	929	M929TEV						

944-972

Dennis Dart 9.8SDL3054   Plaxton Pointer   B39F   1995

944	N944CPU	950	N950CPU	956	N956CPU	962	N962CPU	968	N968CPU
945	N945CPU	951	N951CPU	957	N957CPU	963	N963CPU	969	N969CPU
946	N946CPU	952	N952CPU	958	N958CPU	964	N964CPU	970	N970CPU
947	N947CPU	953	N953CPU	959	N959CPU	965	N965CPU	971	N971CPU
948	N948CPU	954	N954CPU	960	N960CPU	966	N966CPU	972	N972CPU
949	N949CPU	955	N955CPU	961	N961CPU	967	N967CPU		

1001	H101KVX	Leyland Olympian ON2R50C13Z4	Leyland	H47/31F	1990
1002	H102KVX	Leyland Olympian ON2R50C13Z4	Leyland	H47/31F	1990
1003	H103KVX	Leyland Olympian ON2R50C13Z4	Leyland	H47/31F	1990
1004	H104KVX	Leyland Olympian ON2R50C13Z4	Leyland	H47/31F	1990

1400-1424

Leyland Lynx LX112L10ZR1R   Leyland Lynx   B49F   1988   Ex Eastern National, 1990

1400	E400HWC	1409	F409LTW	1412	F412MNO	1419	F419MWC	1422	F422MJN
1404	F404LTW	1410	F410MNO	1417	F417MWC	1420	F420MJN	1423	F423MJN
1405	F405LTW	1411	F411MNO	1418	F418MWC	1421	F421MJN	1424	F424MJN
1406	F406LTW								

1601	L601MWC	Volvo B6-9.9M	Northern Counties Paladin	B40F	1993

1803-1871

Leyland National 11351A/1R   B49F   1977-78 Ex Eastern National, 1990

1803	TJN502R	1830	VNO732S	1845	WJN565S	1853	YEV311S	1868	YEV326S
1822	VNO745S	1839	WJN559S	1846	WJN566S	1857	YEV315S	1869	YEV327S
1823	VAR900S	1840	WJN560S	1847	YEV305S	1864	YEV322S	1871	YEV329S

1876-1920

Leyland National 11351A/1R   B49F   1978-79 Ex Eastern National, 1990

1876	BNO666T	1887	BNO677T	1896	DAR118T	1907	DAR129T	1913	JHJ139V
1878	BNO668T	1889	BNO679T	1897	DAR119T	1908	DAR130T	1915	JHJ141V
1881	BNO671T	1891w	BNO681T	1898	DAR120T	1909	DAR131T	1918	JHJ144V
1882	BNO672T	1894	BNO684T	1903	DAR125T	1911	DAR133T	1919	JHJ145V
1884	BNO674T	1895	BNO685T	1906	DAR128T	1912	DAR134T	1920w	JHJ146V
1886	BNO676T								

2202w	FDZ984	Leyland Tiger TRCTL11/3R	Duple Goldliner	C48FT	1982	Ex Western National, 1994
3110	XHK215X	Bristol VRT/SL3/6LXB	Eastern Coach Works	H43/31F	1981	Ex Eastern National, 1990
3113	XHK218X	Bristol VRT/SL3/6LXB	Eastern Coach Works	H43/31F	1981	Ex Eastern National, 1990
4000	XHK235X	Leyland Olympian ONLXB/1R	Eastern Coach Works	H45/32F	1981	Ex Eastern National, 1990
4001	XHK236X	Leyland Olympian ONLXB/1R	Eastern Coach Works	H45/32F	1981	Ex Eastern National, 1990
4002	XHK237X	Leyland Olympian ONLXB/1R	Eastern Coach Works	H45/32F	1981	Ex Eastern National, 1990
4003	B698BPU	Leyland Olympian ONLXB/1R	Eastern Coach Works	H45/32F	1984	Ex Eastern National, 1991
4004	B699BPU	Leyland Olympian ONLXB/1R	Eastern Coach Works	H45/32F	1984	Ex Eastern National, 1992
4006	C712GEV	Leyland Olympian ONLXB/1R	Eastern Coach Works	H45/32F	1985	Ex Eastern National, 1991
4009	C409HJN	Leyland Olympian ONLXB/1RH	Eastern Coach Works	DPH42/30F	1985	Ex Eastern National, 1990

Livery: Canary yellow and purple; Yellow, orange and blue (City Saver) 511-22, 601-19

Previous Registrations:

ATH110T   AFH192T, MKH487A, AEP253T, 999BCY, ATH58T, 278TNY   FDZ984   OHM831Y

WESSEX

Wessex National Ltd, Premier House, Sussex Street, St Phillips, Bristol, BS2 0RB

130	K461PNR	Volvo B10M-60	Plaxton Premiére 350	C51FT	1993	
131	L65UOU	Volvo B10M-60	Plaxton Premiére 350	C49FT	1994	
138	K65OHT	Volvo B10M-60	Plaxton Premiére 350	C49FT	1993	
139	K67OHT	Volvo B10M-60	Plaxton Premiére 350	C49FT	1993	
140	H68PDW	Volvo B10M-60	Plaxton Paramount 3500 III	C49FT	1991	Ex Bebb, Llantwit, Fardre, 1993
141	H69PDW	Volvo B10M-60	Plaxton Paramount 3500 III	C49FT	1991	Ex Bebb, Llantwit, Fardre, 1993
142	H71PWO	Volvo B10M-60	Plaxton Paramount 3500 III	C49FT	1991	Ex Bebb, Llantwit, Fardre, 1993
144w	E944LAE	Iveco Daily 49.10	Robin Hood City Nippy	C17F	1988	Ex City Line, 1991
145	H201JHP	Peugeot-Talbot Pullman	Talbot	B22F	1990	Ex Midland Red West, 1995

149-158

Volvo B10M-60 — Plaxton Expressliner — C46FT* — 1989-91 *149-53 are C49FT

149	H461BEU	152	H194BTC	154	J429GHT	156	J431GHT	158	J204HWS
150	H462BEU	151	H193BTC	153	H195BTC	155	J430GHT	157	J203HWS

159-163

Volvo B10M-60 — Plaxton Expressliner 2 — C46FT — 1993

159	K509NOU	160	K991OEU	161	K792OTC	162	K793OTC	163	K794OTC

165	G389PNV	Volvo B10M-60	Plaxton Expressliner	C49FT	1990	Ex Premier Travel, 1993
166	L64UOU	Volvo B10M-62	Plaxton Expressliner 2	C49FT	1993	
167	L67UOU	Volvo B10M-60	Plaxton Expressliner 2	C49FT	1993	
169	M92BOU	Volvo B10M-62	Plaxton Expressliner 2	C48FT	1994	
170	G388PNV	Volvo B10M-60	Plaxton Expressliner	C46FT	1990	Ex Express Travel, 1994
171	H72PWO	Volvo B10M-60	Plaxton Expressliner	C49FT	1991	Ex Express Travel, 1994
172	G971KTX	Volvo B10M-60	Plaxton Expressliner	C46FT	1990	Ex Express Travel, 1994
173	G972KTX	Volvo B10M-60	Plaxton Expressliner	C46FT	1990	Ex Express Travel, 1994

Sky Blue is the operating name of Wessex local services which employs almost solely Leyland Nationals. Seen in Bristol is 701, JJG885P, an example transferred from Brewers in 1993 and recently sold. *Phillip Stephenson*

While most of the Wessex fleet carry National Express livery there are a few with the attractive Wessex gold and red scheme. Photographed while working a tour for Cosmos is 130, K461PNR, a Volvo B10M with Plaxton Premiére 350 bodywork. *R A Smith*

174	M763CWS	Volvo B10M-62	Plaxton Expressliner 2	C46FT	1994	
175	M764CWS	Volvo B10M-62	Plaxton Expressliner 2	C46FT	1994	
176	M765CWS	Volvo B10M-62	Plaxton Expressliner 2	C46FT	1994	
177	M413DEV	Volvo B10M-62	Plaxton Expressliner 2	C49FT	1995	
178	M439FHW	Volvo B10M-62	Plaxton Expressliner 2	C46FT	1995	
179	M440FHW	Volvo B10M-62	Plaxton Expressliner 2	C46FT	1995	
180	M41FTC	Volvo B10M-62	Plaxton Expressliner 2	C46FT	1995	

Sky Bue Buses:

703	NWS906R	Leyland National 11351A/2R		B49F	1977	Ex Brewers, 1993
704	PEV690R	Leyland National 11351A/1R		B49F	1976	Ex Brewers, 1993
705	D756RWC	Ford Transit VE6	Dormobile	B16F	1986	Ex Thamesway, 1993

707-713 Leyland National 11351A/1R B52F 1976-79 Ex Badgerline, 1994

707	NFB599R	**710**	OHW489R	**711**	SAE757S	**712**	TTC537T	**713**	YFB969V
708	NFB601R								

Livery: Silver, gold and red (Wessex private hire) 130/1/9; blue (Sky Blue Buses) 701-713; silver, blue, black and red (Air Link) 159-63/6/7/9/74-9; white (Eurolines) 138; National Express, remainder.

WESTERN NATIONAL

Western National Ltd, Western House, 38 Lemon Street, Truro, Cornwall, TR1 2NS

Depots and major outstations: Union Street, Camborne; New Road, Callington; Little Cotton Farm, Dartmouth; Tregonnigie Industrial Estate, Falmouth; Flambards Car Park, Helston; Tolcarne Street, Newquay; Trecerus Industrial Estate, Padstow; Long Rock Industrial Estate, Penzance; Laira Bridge Road, Plymouth; Elliot Road, St Austell; Crowndale Road, Tavistock; Trevol Road, Torpoint; Wills Road Industrial Estate, Totnes and Lemon Quay, Truro. **Other outstations:** Bodmin, Canworthy Water, Delabole, Launceston, Liskard, North Petherwin, Pelynt, St Just, Splatt, Tregony.

40	C201PCD	Mercedes-Benz L608D	Alexander AM	B20F	1986	Ex Brighton & Hove, 1990
41	C202PCD	Mercedes-Benz L608D	Alexander AM	B20F	1986	Ex Brighton & Hove, 1990
42	C211PCD	Mercedes-Benz L608D	Alexander AM	B20F	1986	Ex Brighton & Hove, 1990
51	B38AAF	Mercedes-Benz L608D	Reeve Burgess	B16FL	1984	
54	B41AAF	Mercedes-Benz L608D	Reeve Burgess	B16FL	1984	
56	B43AAF	Mercedes-Benz L608D	Reeve Burgess	B19FL	1984	

57-79

Mercedes-Benz L608D — Reeve Burgess — B20F — 1985

57	C672ECV	60	C675ECV	64	C679ECV	71	C689ECV	76	C691ECV
58	C673ECV	61	C676ECV	67	C682ECV	74	C686ECV	78	C693ECV
59	C674ECV	62	C677ECV	68	C683ECV	75	C690ECV	79	C694ECV

86-125

Mercedes-Benz L608D — Reeve Burgess — B20F* — 1985-86 *119/22/4/5 are DP19F

86	C783FRL	95	C792FRL	102	C799FRL	108	C951GAF	115	C958GAF
87	C784FRL	96	C793FRL	103	C800FRL	109	C952GAF	116	C959GAF
88	C785FRL	97	C794FRL	104	C801FRL	110	C953GAF	119	C965GCV
89	C786FRL	98	C795FRL	105	C802FRL	111	C954GAF	122	C968GCV
90	C787FRL	99	C796FRL	106	C949GAF	112	C955GAF	124	C970GCV
92	C789FRL	100	C797FRL	107	C950GAF	113	C956GAF	125	C489HCV
94	C791FRL	101	C798FRL						

140-156

Mercedes-Benz L608D — Reeve Burgess — B20F — 1986

140	C229HCV	147	C100HGL	150	C103HGL	152	C105HGL	155	C108HGL
145	C98HGL	148	C101HGL	151	C104HGL	153	C106HGL	156	C109HGL
146	C99HGL	149	C102HGL						

158-174

Mercedes-Benz L608D — Reeve Burgess — B20F — 1986 — Ex Eastern National, 1994

158	C208HJN	164	C214HJN	168	C218HJN	170	C220HJN	174	C224HJN
163	C213HJN	166	C216HJN	169	C219HJN	173	C223HJN		

200-205

Leyland Lynx LX112TL11R1R — Leyland Lynx — B51F — 1988

200	E200BOD	202	E202BOD	203	E203BOD	204	E204BOD	205	E205BOD
201	E201BOD								

221	M221EAF	Dennis Lance SLF 11SDA3201	Wright Pathfinder 320	B40F	1995

301-326

Mercedes-Benz 811D — Carlyle — B31F* — 1990-91 *301/2/11 are DP31F 325/6 are B33F

301	G151GOL	307	H892LOX	312	H712HGL	317	H717HGL	322	H722HGL
302	G152GOL	308	H893LOX	313	H713HGL	318	H718HGL	323	H723HGL
303	G153GOL	309	H894LOX	314	H714HGL	319	H719HGL	324	H724HGL
304	G154GOL	310	H895LOX	315	H715HGL	320	H720HGL	325	H725HGL
305	G155GOL	311	H896LOX	316	H716HGL	321	H721HGL	326	H726HGL
306	H891LOX								

Opposite, top: **Currently Western National's only Dennis Lance is also the first SLF-Super Low Floor in the South Western fleets. It is seen fully lettered for its dedicated service, 183 from Plymouth to Okehampton.** *Malc McDonald*

Opposite, bottom: **The narrow streets of the picturesque Cornish fishing port of Mevagissey play host to Western National 104, C801FRL, one of the large number of Mercedes-Benz L608Ds to enter service in the mid 1980s.** *David Donati*

In addition to the standard length of Dennis Darts, three shorter versions joined the Western National fleet in 1995. Seen in Plymouth is 503, M503CCV *Malc McDonald*

327	E807MOU	Mercedes-Benz 811D		Optare StarRider		B31F	1988	Ex Badgerline, 1991	
328	E808MOU	Mercedes-Benz 811D		Optare StarRider		B33F	1988	Ex Badgerline, 1991	
329	E812MOU	Mercedes-Benz 811D		Optare StarRider		B31F	1988	Ex Badgerline, 1991	
330	E820MOU	Mercedes-Benz 811D		Optare StarRider		B31F	1988	Ex Badgerline, 1991	

331-343   Mercedes-Benz 811D   Plaxton Beaver   B31F   1992

331	K331OAF	334	K334OAF	337	K337OAF	340	K340OAF	342	K342OAF
332	K332OAF	335	K335OAF	338	K338OAF	341	K341OAF	343	K343OAF
333	K333OAF	336	K336OAF	339	K339OAF				

344-360   Mercedes-Benz 811D   Plaxton Beaver   B31F   1993-94

344	K344ORL	348	K348ORL	352	K352ORL	355	L355VCV	358	L358VCV
345	K345ORL	349	K349ORL	353	K353ORL	356	L356VCV	359	L359VCV
346	K346ORL	350	K350ORL	354	K354ORL	357	L357VCV	360	L360VCV
347	K347ORL	351	K351ORL						

401-406   Dennis Dart 9.8SDL3035   Plaxton Pointer   B40F   1994

401	L401VCV	403	L403VCV	404	L404VCV	405	L405VCV	406	L406VCV
402	L402VCV								

407-426   Dennis Dart 9.8SDL3054   Plaxton Pointer   B38F   1995   422-6 are DP37F

407	M407CCV	411	M411CCV	415	M415CCV	419	M419CCV	423	M423CCV
408	M408CCV	412	M412CCV	416	M416CCV	420	M420CCV	424	M424CCV
409	M409CCV	413	M413CCV	417	M417CCV	421	M421CCV	425	M425CCV
410	M410CCV	414	M414CCV	418	M418CCV	422	M422CCV	426	M426CCV

501	M501CCV	Dennis Dart 9SDL3053	Plaxton Pointer	B35F	1995	
502	M502CCV	Dennis Dart 9SDL3053	Plaxton Pointer	B35F	1995	
503	M503CCV	Dennis Dart 9SDL3053	Plaxton Pointer	B35F	1995	
518	D44OYA	Fiat 35.8	G&M	C14F	1987	Ex Robert's, Plympton, 1988
520	H324HVT	Mercedes-Benz 609D	PMT	C24F	1990	Ex Plymouth Argyle AFC, 1994
524	E683XVU	Ford Transit 130	Deansgate	M12	1988	

The sparse population of the Cornish peninsula has required small capacity vehicles to be used over many years. Mercedes-Benz 709D 625, K625ORL, with Plaxton Beaver bodywork is the 1990s offering to meet this need. *David Donati collection*

525	F999UGL	Ford Transit 130		Deansgate		M12	1988			
526	F21URL	Ford Transit 130		Deansgate		M12	1988			
550	F22URL	Ford Transit 130		Deansgate		M12	1988			
553	LFR293X	Iveco 79F.10		Harwin		C29F	1982	Ex Robert's, Plympton, 1988		
557	C159DWT	Fiat 79.14 Caetano Viana		C19F			1986	Ex Robert's, Plympton, 1988		
573	B476YEU	Ford Transit 190		Carlyle		DP16F	1985	Ex Badgerline, 1991		
574	D83KRL	Ford Transit 190		Dormobile		DP16F	1986	Ex Badgerline, 1995		
580	G860VAY	Toyota Coaster HB31R		Caetano Optimo		C21F	1989	Ex Yeates, 1994		
581	J727KBC	Toyota Coaster HDB30R		Caetano Optimo II		C18F	1991	Ex Spirit of London, 1994		
582	M582DAF	Toyota Coaster HZB50R		Caetano Optimo III		C21F	1995			

601-625

Mercedes-Benz 709D · Plaxton Beaver · B23F* · 1993 · *620-5 are B25F

601	K601ORL	606	K606ORL	611	K611ORL	616	K616ORL	621	K621ORL
602	K602ORL	607	K607ORL	612	K612ORL	617	K617ORL	622	K622ORL
603	K603ORL	608	K608ORL	613	K613ORL	618	K618ORL	623	K623ORL
604	K604ORL	609	K609ORL	614	K614ORL	619	K619ORL	624	K624ORL
605	K605ORL	610	K610ORL	615	K615ORL	620	K620ORL	625	K625ORL

626	H825ERV	Mercedes-Benz 709D	Wadham Stringer Wessex	B25F	1991	Ex Roselyn Coaches, Par, 1993
627	J901MAF	Mercedes-Benz 709D	Wadham Stringer Wessex	B21F	1991	Ex Roselyn Coaches, Par, 1993

628-651

Mercedes-Benz 709D · Plaxton Beaver · B23F · 1994

628	L628VCV	633	L633VCV	638	L638VCV	643	L643VCV	648	L648VCV
629	L629VCV	634	L634VCV	639	L639VCV	644	L644VCV	649	L649VCV
630	L630VCV	635	L635VCV	640	L640VCV	645	L645VCV	650	L650VCV
631	L631VCV	636	L636VCV	641	L641VCV	646	L646VCV	651	L651VCV
632	L632VCV	637	L637VCV	642	L642VCV	647	L647VCV		

750-757

Leyland Olympian ONLXB/1R · Eastern Coach Works · DPH44/32F · 1983

750	A750VAF	752	A752VAF	754	A754VAF	756	A756VAF	757	A757VAF
751	A751VAF	753	A753VAF	755	A755VAF				

Two convertible open top buses, both Bristol VRTs, remain with Western National. Photographed in Plymouth is 944, VDV144S. *Malc McDonald*

801	K801ORL	Volvo Olympian YN2RV18Z4	Northern Counties Palatine	DPH39/30F	1993	
802	K802ORL	Volvo Olympian YN2RV18Z4	Northern Counties Palatine	DPH39/30F	1993	
803	K803ORL	Volvo Olympian YN2RV18Z4	Northern Counties Palatine	DPH39/30F	1993	
804	K804ORL	Volvo Olympian YN2RV18Z4	Northern Counties Palatine	DPH39/30F	1993	
941	VDV141S	Bristol VRT/SL3/6LXB	Eastern Coach Works	CO43/31F	1978	
944	VDV144S	Bristol VRT/SL3/6LXB	Eastern Coach Works	CO43/31F	1978	
1083	JNU138N	Bristol VRT/SL2/6LX(6LXB)	Eastern Coach Works	H39/31F	1975	Ex Badgerline, 1988
1084	HTC728N	Bristol VRT/SL2/6LX	Eastern Coach Works	H39/31F	1975	Ex Badgerline, 1988
1087	BEP968V	Bristol VRT/SL3/501(6LXB)	Eastern Coach Works	H43/31F	1979	Ex South Wales, 1989
1100	SFJ100R	Bristol VRT/SL3/6LXB	Eastern Coach Works	H43/31F	1977	
1105	SFJ105R	Bristol VRT/SL3/6LXB	Eastern Coach Works	H43/31F	1977	Ex Devon General, 1987
1106	SFJ106R	Bristol VRT/SL3/6LXB	Eastern Coach Works	H43/31F	1977	
1108	VDV108S	Bristol VRT/SL3/6LXB	Eastern Coach Works	H43/31F	1977	
1109	UTO832S	Bristol VRT/SL3/501(6LXB)	Eastern Coach Works	H43/31F	1977	Ex Devon General, 1987

1114-1131

Bristol VRT/SL3/6LXB   Eastern Coach Works   H43/31F   1977-78

1114	VDV114S	1118	VDV118S	1120	VDV120S	1123	XDV603S	1129	XDV609S
1116	VDV116S	1119	VDV119S	1121	VDV121S	1128	XDV608S	1131	XDV601S
1117	VDV117S								

1132-1226

Bristol VRT/SL3/6LXB   Eastern Coach Works   H43/31F   1978-81 1224 is VRT/SL3/6LXC(6LXB)

1132	AFJ697T	1140	AFJ705T	1148	AFJ750T	1176	FDV808V	1201	LFJ845W
1133	AFJ698T	1141	AFJ706T	1149	AFJ751T	1182	FDV814V	1202	LFJ846W
1134	AFJ699T	1142	AFJ744T	1153	AFJ760T	1183	FDV815V	1203	LFJ847W
1135	AFJ700T	1143	AFJ745T	1154	AFJ761T	1197	LFJ841W	1220	LFJ867W
1136	AFJ701T	1144	AFJ746T	1155	AFJ762T	1198	LFJ842W	1224	LFJ871W
1137	AFJ702T	1145	AFJ747T	1174	FDV806V	1199	LFJ843W	1225	LFJ872W
1138	AFJ703T	1147	AFJ749T	1175	FDV807V	1200	LFJ844W	1226	LFJ873W
1139	AFJ704T								

1227	EWS747W	Bristol VRT/SL3/680(6LXB)	Eastern Coach Works	DPH43/31F	1981	Ex Badgerline, 1990

Newquay is one of the larger Cornish towns and it was here that Bristol VRT 1133, AFJ698T was seen. The standard Eastern Coach Works body is typical of many double-deck buses most of which are used to their capacity on school services. *David Donati*

1228-1237

		Bristol VRT/SL3/501*		Eastern Coach Works		H43/31F	1978-79 Ex South Wales, 1990		
							*1228/31/4/5 are VRT/SL3/501(6LXB)		
1228	RTH929S	1230	VTH942T	1232	WTH945T	1234	WTH950T	1236	WTH961T
1229	TWN936S	1231	WTH943T	1233	WTH946T	1235	WTH951T	1237	BEP966V

1238-1251

		Bristol VRT/SL3/6LXB		Eastern Coach Works		H43/31F	1980-81 Ex Thamesway, 1991-92		
1238	KOO785V	1241	XHK231X	1244	UAR589W	1248	UAR590W	1250	XHK220X
1239	UAR595W	1242	XHK228X	1245	UAR594W	1249	XHK223X	1251	XHK230X
1240	XHK225X	1243	UAR586W	1246	UAR597W					

1252	PHY697S	Bristol VRT/SL3/6LXB	Eastern Coach Works	H43/31F	1977	Ex City Line, 1994
1253	TWS915T	Bristol VRT/SL3/6LXB	Eastern Coach Works	H43/31F	1979	Ex City Line, 1994
1254	AHU516V	Bristol VRT/SL3/6LXB	Eastern Coach Works	H43/31F	1980	Ex City Line, 1994
2101	M101ECV	Volvo B12T	Van Hool	CH57/14CT	1995	
2102	M102ECV	Volvo B12T	Van Hool	CH57/14CT	1995	
2103	M103ECV	Volvo B12T	Van Hool	CH57/14CT	1995	
2200	URL94Y	Leyland Tiger TRCTL11/3R	Duple Goldliner	C51F	1982	Ex United Welsh Coaches, 1990
2203	A749VAF	Leyland Tiger TRCTL11/3R	Plaxton Paramount 3500	C51F	1983	
2206	A530WRL	Leyland Tiger TRCTL11/3R	Plaxton Paramount 3200	C53F	1984	
2208	A532WRL	Leyland Tiger TRCTL11/3R	Plaxton Paramount 3200	C53F	1984	
2217	B194BAF	Leyland Tiger TRCTL11/3RH	Plaxton Paramount 3200 II	C46FT	1985	
2218	B195BAF	Leyland Tiger TRCTL11/3RH	Plaxton Paramount 3200 II	C46FT	1985	
2219	B196BAF	Leyland Tiger TRCTL11/3RH	Plaxton Paramount 3200 II	C46FT	1985	
2220	B197BAF	Leyland Tiger TRCTL11/3RH	Plaxton Paramount 3200 II	C46FT	1985	
2239	A747JAY	Volvo B10M-61	Duple Caribbean	C49FT	1984	Ex Grenville Motors, Troon, 1988
2243	J243LGL	Volvo B10M-60	Plaxton Expressliner	C46FT	1992	
2244	J244LGL	Volvo B10M-60	Plaxton Expressliner	C46FT	1992	
2245	J245LGL	Volvo B10M-60	Plaxton Expressliner	C46FT	1992	
2246	J246LGL	Volvo B10M-60	Plaxton Expressliner	C46FT	1992	
2247	F444DUG	Volvo B10M-61	Plaxton Paramount 3500 III	C48FT	1989	Ex Wallace Arnold, 1992
2248	F446DUG	Volvo B10M-61	Plaxton Paramount 3500 III	C48FT	1989	Ex Wallace Arnold, 1992
2249	K249PCV	Volvo B10M-60	Plaxton Expressliner 2	C46FT	1993	
2250	K250PCV	Volvo B10M-60	Plaxton Expressliner 2	C46FT	1993	
2251	K251PCV	Volvo B10M-60	Plaxton Expressliner 2	C46FT	1993	

2252	F445DUG	Volvo B10M-60	Plaxton Paramount 3500 III	C51FT	1989	Ex Wallace Arnold, 1993
2253	G541LWU	Volvo B10M-60	Plaxton Paramount 3500 III	C51FT	1990	Ex Wallace Arnold, 1993
2254	L254UCV	Volvo B10M-60	Plaxton Expressliner 2	C46FT	1993	
2255	L255UCV	Volvo B10M-60	Plaxton Expressliner 2	C46FT	1993	
2256	L256UCV	Volvo B10M-60	Plaxton Expressliner 2	C46FT	1993	
2257	L257UCV	Volvo B10M-60	Plaxton Expressliner 2	C46FT	1993	
2258	H613UWR	Volvo B10M-60	Plaxton Paramount 3500 III	C46FT	1991	Ex Wallace Arnold, 1994
2259	H614UWR	Volvo B10M-60	Plaxton Paramount 3500 III	C46FT	1991	Ex Wallace Arnold, 1994
2260	H615UWR	Volvo B10M-60	Plaxton Paramount 3500 III	C46FT	1991	Ex Wallace Arnold, 1994
2261	FDZ980	Volvo B10M-60	Plaxton Paramount 3500 III	C49FT	1990	Ex Wallace Arnold, 1994
2301	M301BRL	Volvo B10M-62	Plaxton Expressliner 2	C46FT	1994	
2302	M302BRL	Volvo B10M-62	Plaxton Expressliner 2	C46FT	1994	
2303	M303BRL	Volvo B10M-62	Plaxton Expressliner 2	C46FT	1994	
2401	J701CWT	Volvo B10M-60	Plaxton Première 350	C48FT	1992	Ex Wallace Arnold, 1995
2402	J703CWT	Volvo B10M-60	Plaxton Première 350	C46FT	1992	Ex Wallace Arnold, 1995
2504	ODT232	Volvo B10M-61	Van Hool Alizée	C48FT	1987	Ex Badgerline, 1995
2505	XFF283	Volvo B10M-61	Van Hool Alizée	C48FT	1987	Ex Badgerline, 1995
2506	EWV665	Volvo B10M-61	Van Hool Alizée	C48FT	1987	Ex Badgerline, 1995
2507	RUH346	Volvo B10M-61	Van Hool Alizée	C48FT	1987	Ex Badgerline, 1995
2508	UWB183	Volvo B10M-61	Van Hool Alizée	C53F	1987	Ex Badgerline, 1991
2509	OWB243	Volvo B10M-61	Van Hool Alizée	C48FT	1987	Ex Badgerline, 1991
2510	PSU527	Volvo B10M-61	Van Hool Alizée	C48FT	1987	Ex Badgerline, 1993
2511	UHW661	Volvo B10M-61	Van Hool Alizée	C48FT	1987	Ex Badgerline, 1991
3425	GTA806N	Leyland Leopard PSU3B/4R	Plaxton Supreme III(1977)	C51F	1975	Ex Devon General, 1988
3440	SFJ140R	Leyland Leopard PSU3E/4RT	Plaxton Supreme III	C47F	1977	
3444	SFJ144R	Leyland Leopard PSU3D/4RT	Plaxton Supreme III	C47F	1977	
3450	SFJ150R	Leyland Leopard PSU3E/4R	Plaxton Supreme III	C47F	1977	
3453	CPT823S	Leyland Leopard PSU3E/4R	Plaxton Supreme III Express	C53F	1978	Ex Grenville Motors, Troon, 1987
3500	KTT808P	Leyland Leopard PSU3B/4R	Plaxton Supreme III Express	DP53F	1975	
3508	SFJ158R	Leyland Leopard PSU3C/4R	Plaxton Supreme III Express	C51F	1977	
3514	URL992S	Leyland Leopard PSU3E/4R	Plaxton Supreme III Express	C49F	1978	Ex Thomas, Relubbus, 1995
3515	ETH68V	Leyland Leopard PSU3E/4R	Plaxton Supreme IV Express	C49F	1980	Ex Brewers, 1995

3516-3524

		Leyland Leopard PSU3E/4R(DAF) Plaxton Supreme III Express C49F				1978-79		
3516	VOD616S	3518	VOD618S	3522	AFJ714T	3523	AFJ715T	3524	AFJ716T
3517	VOD617S							

3538-3546

		Leyland Leopard PSU3E/4R(DAF) Plaxton Supreme IV Express C49F				1980		
3538	FDV794V	3542	FDV798V	3544	FDV800V	3545	FDV801V	3546	FDV802V

Eight Leyland Olympians with Eastern Coach Works bodies were delivered to Western National in 1983 while it remained part of the National Bus Company. They feature high-back seats for the longer limited stop services such as the 93 from Dartmouth to Plymouth through Kingsbridge. At the end of its journey in Plymouth is 750, A750VAF.
Malc McDonald

Re-engined with a DAF unit, Leyland Leopard 3542, FDV798V is seen returning to Newquay depot after operating a school service. The body type is an express version of the Plaxton Supreme IV, and one of six of the type now refurbished. *David Donati*

3549	FDV805V	Leyland Leopard PSU3F/5R(DAF)	Plaxton Supreme IV Express	C49F	1980	
3550	JTH44W	Leyland Leopard PSU3F/5R	Plaxton Supreme IV Express	C53F	1982	Ex Brewers, 1994
3551	NTH156X	Leyland Leopard PSU3F/5RT	Duple Dominant IV Express	C53F	1981	Ex Brewers, 1994
3553	JWE243W	Leyland Leopard PSU5D/4R	Plaxton Supreme IV	C46FT	1980	Ex Brewers, 1995
3555	JUP113T	Leyland Leopard PSU5C/4R	Plaxton Supreme III	C51F	1979	Ex Thomas, Relubbus, 1995
3580	VJT738	Leyland Leopard PSU3E/4R	Plaxton P'mount 3200E(1983)	C49F	1977	Ex Brewers, 1995
3591	ROG550Y	Leyland Leopard PSU3A/4R	Eastern Coach Works (1983)	C49F	1970	Ex Brewers, 1995
3592	ANA92Y	Leyland Leopard PSU5E/4R	Eastern Coach Works B51	C53F	1983	Ex Brewers, 1995
3593	ANA93Y	Leyland Leopard PSU5E/4R	Eastern Coach Works B51	C53F	1982	Ex Brewers, 1995
3594	ANA94Y	Leyland Leopard PSU5E/4R	Eastern Coach Works B51	C53F	1983	Ex Brewers, 1995
7042	D617YCX	DAF MB230DKVL615	Van Hool Alizée	C53FT	1987	Ex Lidgey, Tregony, 1995
7043	F767XNH	LAG G355Z	LAG Panoramic	C49FT	1989	Ex Lidgey, Tregony, 1995

Previous Registrations:

A992SGK	AEH632Y	ODT232	D504GHY	URL94Y	OHM832Y, FDZ980
ETH68V	KUB669V, 278TNY	OWB243	D509HHW	URL991S	VPH35S, OO1961
EWV665	D506GHY	PSU527	D510HHW	URL992S	UWE89S, GUN162
FDZ980	G521LWU	ROG550Y	WHA251H	UWB183	D508HHW
J901MAF	J6EDE	RUH346	D507GHY	VJT738	PWS492S
JTH44W	GTH536W, 948RJO	UHW661	D551HHW	XFF283	D505GHY
NTH156X	MEP969X, YBK132				

Livery: White, blue and red; (Robert's Coaches): 520/57/80-2, National Express: 2101-3, 2243-51/4-60, 2301-3, 2402.
Named vehicles: 344 *Tegen*, 345 *Kerensa*, 346 *Cordelia*, 347 *Demelza*, 348 *Lowenna*.

Vehicle index

Reg	Operator	Reg	Operator	Reg	Operator	Reg	Operator
2GRT	Grampian	A105EBC	Durbin Coaches	A546KUM	Rider Group	A949SAE	City Line
23PTA	Rider Group	A105FSA	Grampian	A547KUM	Rider Group	A950SAE	City Line
101ASV	Midland Bluebird	A105KUM	Rider Group	A549KUM	Rider Group	A951SAE	City Line
110ASV	Midland Bluebird	A106FSA	Grampian	A599NYG	Rider Group	A952SAE	City Line
119ASV	Midland Bluebird	A106KUM	Rider Group	A600NYG	Rider Group	A953SAE	City Line
131ASV	Mair's	A107FSA	Grampian	A601NYG	Rider Group	A954SAE	City Line
143ASV	Midland Bluebird	A107KUM	Rider Group	A611XKU	Durbin Coaches	A955THW	City Line
144ASV	Midland Bluebird	A108FSA	Grampian	A656VDA	Midland Red West	A956THW	City Line
156ASV	Midland Bluebird	A108KUM	Rider Group	A657VDA	Midland Red West	A957THW	City Line
158ASV	Lowland	A109FSA	Grampian	A658KUM	Rider Group	A958THW	City Line
278TNY	Brewers	A109KUM	Rider Group	A658VDA	Midland Red West	A959THW	City Line
300CUH	Brewers	A110FSA	Grampian	A659KUM	Rider Group	A960THW	City Line
365UMY	Midland Bluebird	A110KUM	Rider Group	A660KUM	Eastern National	A961THW	City Line
373GRT	Midland Bluebird	A111KUM	Rider Group	A661KUM	Rider Group	A962THW	City Line
507EXA	PMT	A112KUM	Rider Group	A662KUM	Rider Group	A963THW	City Line
542GRT	Leicester	A113KUM	Rider Group	A663KUM	Eastern National	A964THW	City Line
605BBO	Brewers	A114KUM	Rider Group	A664KUM	Rider Group	A965THW	City Line
692FFC	Midland Bluebird	A115KUM	Rider Group	A665KUM	Eastern National	A966THW	City Line
693AFU	Midland Bluebird	A116KUM	Rider Group	A666KUM	Rider Group	A967THW	City Line
737ABD	Mair's	A117KUM	Rider Group	A667KUM	Rider Group	A968THW	City Line
781GRT	Kirkpatrick	A118KUM	Rider Group	A668KUM	Eastern National	AAE645V	Badgerline
948RJO	Brewers	A119KUM	Rider Group	A669KUM	Rider Group	AAE652V	Badgerline
972SYD	Brewers	A120KUM	Rider Group	A669XDA	Midland Red West	AAE653V	Badgerline
6149KP	Eastern Counties	A121KUM	Rider Group	A670KUM	Rider Group	AAE654V	Badgerline
6920MX	Eastern Counties	A136BSC	SMT	A670XUK	Midland Red West	AAE655V	Badgerline
7694VC	Eastern Counties	A136SMA	PMT	A671KUM	Rider Group	AAE656V	Badgerline
7881UA	Midland Bluebird	A137BSC	SMT	A672KUM	Rider Group	AAE657V	Badgerline
8995WY	Rider Group	A137SMA	PMT	A678KDV	Midland Red West	AAE661V	Badgerline
A9SMT	SMT	A138BSC	SMT	A686MWX	Rider Group	AAE662V	Badgerline
A10SMT	SMT	A138SMA	PMT	A691OHJ	Brewers	AAE663V	Badgerline
A12SMT	SMT	A139BSC	SMT	A692OHJ	Eastern Counties	AAE664V	Badgerline
A13SMT	SMT	A140BSC	SMT	A693OHJ	Brewers	ABD74X	Northampton
A15SMT	SMT	A141BSC	SMT	A695OHJ	Brewers	ABD75X	Northampton
A16SMT	SMT	A142BSC	SMT	A733GFA	PMT	ABD76X	Northampton
A17SMT	SMT	A143BSC	Lowland	A734GFA	PMT	ABR868S	Rider Group
A18SMT	SMT	A143SMA	PMT	A735GFA	PMT	ABR869S	Rider Group
A19SMT	SMT	A144SMA	PMT	A736GFA	PMT	ACY178A	SWT
A20SMT	SMT	A145SMA	PMT	A737GFA	PMT	AFJ697T	Western National
A71FRY	Leicester	A146UDM	PMT	A738GFA	PMT	AFJ698T	Western National
A72FRY	Leicester	A156UDM	PMT	A739GFA	PMT	AFJ699T	Western National
A73FRY	Leicester	A157UDM	PMT	A740GFA	PMT	AFJ700T	Western National
A74FRY	Leicester	A158UDM	PMT	A741GFA	PMT	AFJ701T	Western National
A75FRY	Leicester	A159UDM	PMT	A742GFA	PMT	AFJ702T	Western National
A76FRY	Leicester	A160UDM	PMT	A743JRE	PMT	AFJ703T	Western National
A77FRY	SMT	A161VDM	PMT	A744JRE	PMT	AFJ704T	Western National
A77RRP	Leicester	A162VDM	PMT	A745JRE	PMT	AFJ705T	Western National
A78FRY	Leicester	A163VDM	PMT	A746JRE	PMT	AFJ706T	Western National
A78RRP	SMT	A164VDM	PMT	A747JAY	Western National	AFJ714T	Western National
A79RRP	SMT	A165VDM	PMT	A747JRE	PMT	AFJ715T	Western National
A80RRP	SMT	A166VFM	PMT	A749VAF	Western National	AFJ716T	Western National
A81RRP	SMT	A167VFM	PMT	A750LWY	Rider Group	AFJ744T	Western National
A82KUM	Rider Group	A168VFM	PMT	A750VAF	Western National	AFJ745T	Western National
A82RRP	SMT	A169VFM	PMT	A751LWY	Rider Group	AFJ746T	Western National
A83KUM	Rider Group	A170VFM	PMT	A751VAF	Western National	AFJ747T	Western National
A84KUM	Rider Group	A171VFM	PMT	A752LWY	Rider Group	AFJ748T	People's Provincial
A85KUM	Rider Group	A201YWP	Badgerline	A752VAF	Western National	AFJ749T	Western National
A86KUM	Rider Group	A202YWP	Badgerline	A753LWY	Rider Group	AFJ750T	Western National
A87KUM	Rider Group	A203YWP	Badgerline	A753VAF	Western National	AFJ751T	Western National
A88KUM	Rider Group	A204YWP	Badgerline	A754LWY	Rider Group	AFJ752T	People's Provincial
A89KUM	Rider Group	A205YWP	Badgerline	A754VAF	Western National	AFJ755T	Midland Red West
A90KUM	Rider Group	A301KJT	People's Provincial	A755LWY	Rider Group	AFJ756T	Midland Red West
A91KUM	Rider Group	A302KJT	People's Provincial	A755VAF	Western National	AFJ760T	Western National
A92KUM	Rider Group	A322BSC	Lowland	A756LWY	Rider Group	AFJ761T	Western National
A93KUM	Rider Group	A323BSC	Lowland	A756VAF	Western National	AFJ762T	Western National
A94KUM	Rider Group	A324BSC	Lowland	A757VAF	Western National	AFJ763T	People's Provincial
A95KUM	Rider Group	A325BSC	Lowland	A758LWY	Rider Group	AHU514V	People's Provincial
A96KUM	Rider Group	A326BSC	Lowland	A760LWY	Rider Group	AHU515V	PMT
A97KUM	Rider Group	A327BSC	Lowland	A809THW	Badgerline	AHU516V	Western National
A98KUM	City Line	A328BSC	SMT	A810THW	Badgerline	AHU517V	City Line
A99KUM	City Line	A329BSC	SMT	A811THW	Badgerline	AHU523V	Badgerline
A100KUM	Rider Group	A455JJF	Durbin Coaches	A812THW	Badgerline	AHW198V	City Line
A101FSA	Grampian	A470GMS	Midland Bluebird	A813THW	Badgerline	AHW200V	City Line
A101KUM	Rider Group	A477GMS	Midland Bluebird	A814THW	Badgerline	AHW201V	Durbin Coaches
A102FSA	Grampian	A530WRL	Western National	A895KCL	Midland Red West	AHW202V	City Line
A102KUM	Rider Group	A532WRL	Western National	A896KCL	Midland Red West	AHW203V	PMT
A103FSA	Grampian	A541KUM	Rider Group	A945SAE	City Line	ALS116Y	Lowland
A103KUM	Rider Group	A542KUM	Rider Group	A946SAE	City Line	ALS117Y	SMT
A104FSA	Grampian	A544KUM	Rider Group	A947SAE	City Line	ALS118Y	SMT
A104KUM	Rider Group	A545KUM	Rider Group	A948SAE	City Line	ALS119Y	SMT

Reg	Operator	Reg	Operator	Reg	Operator	Reg	Operator
ALS122Y	SMT	B119MSO	Grampian	B563RWY	Rider Group	BNO681T	Thamesway
ALS123Y	SMT	B119RRE	PMT	B564RWY	Rider Group	BNO684T	Thamesway
ALS124Y	SMT	B120MSO	Grampian	B565RWY	Rider Group	BNO685T	Thamesway
ALS125Y	SMT	B121MSO	Grampian	B566BOK	Midland Red West	BOU1V	Badgerline
ALS126Y	SMT	B122RWY	Rider Group	B566RWY	Rider Group	BOU2V	Badgerline
ALS127Y	SMT	B123RWY	Rider Group	B567BOK	Midland Red West	BOU3V	Badgerline
ALS128Y	SMT	B124PEL	Rider Group	B567RWY	Rider Group	BOU4V	Badgerline
ALS129Y	SMT	B124RWY	Rider Group	B568BOK	Midland Red West	BOU5V	Badgerline
ALS132Y	SMT	B125RWY	Rider Group	B568LSC	SMT	BOU7V	Badgerline
ALS133Y	SMT	B126RWY	Rider Group	B568RWY	Rider Group	BOU8V	Badgerline
ALS134Y	SMT	B127RWY	Rider Group	B569RWY	Rider Group	BRF689T	PMT
ALS135Y	SMT	B128RWY	Rider Group	B570RWY	Rider Group	BRF691T	Eastern Counties
ANA25T	Rider Group	B129RWY	Rider Group	B571RWY	Rider Group	BRF693T	PMT
ANA31T	Rider Group	B130RWY	Rider Group	B572RWY	Rider Group	BSF765S	Lowland
ANA33T	Rider Group	B131RWY	Rider Group	B575RWY	Rider Group	BSF771S	Lowland
ANA34T	Rider Group	B132RWY	Rider Group	B577RWY	Rider Group	BSS76	Lowland
ANA40T	Rider Group	B133RWY	Rider Group	B578RWY	Rider Group	BSV807	Midland Bluebird
ANA44T	Rider Group	B134RWY	Rider Group	B579RWY	Rider Group	BUT18Y	Leicester
ANA45T	Rider Group	B135RWY	Rider Group	B580RWY	Rider Group	BVG218T	Eastern Counties
ANA48T	Rider Group	B136RWY	Rider Group	B581MLS	Midland Bluebird	BVG219T	Eastern Counties
ANA92Y	Western National	B137RWY	Rider Group	B582MLS	Midland Bluebird	BVG220T	Eastern Counties
ANA93Y	Western National	B138RWY	Rider Group	B583MLS	Midland Bluebird	BVG221T	Eastern Counties
ANA94Y	Western National	B139RWY	Rider Group	B584MLS	Midland Bluebird	BVG222T	Eastern Counties
ANK316X	King's of Dunblain	B140RWY	Rider Group	B585MLS	Midland Bluebird	BVG223T	Eastern Counties
ANO271S	Eastern National	B141RWY	Rider Group	B587MLS	Midland Bluebird	BVG224T	Eastern Counties
ARY225K	Leicester	B142RWY	Rider Group	B588MLS	Midland Bluebird	BVG225T	Eastern Counties
ATH110T	Thamesway	B143RWY	Rider Group	B610VWU	Rider Group	BVP770V	Midland Red West
AUT31Y	Leicester	B144RWY	Rider Group	B689BPU	Eastern National	BVP774V	Midland Red West
AUT32Y	Leicester	B145RWY	Rider Group	B691BPU	Eastern National	BVP775V	Midland Red West
AUT33Y	Leicester	B149GSC	SMT	B696WAR	Eastern National	BVP776V	Midland Red West
AUT34Y	Leicester	B150GSC	SMT	B697WAR	Eastern National	BVP777V	Midland Red West
AUT35Y	Leicester	B151GSC	SMT	B698BPU	Thamesway	BVP778V	Midland Red West
AUT70Y	Leicester	B152GSC	SMT	B699BPU	Thamesway	BVP781V	Midland Red West
AWG623	SMT	B153GSC	SMT	B905DHB	Brewers	BVP782V	Midland Red West
AWN810V	Brewers	B154GSC	SMT	B906DHB	Brewers	BVP783V	Midland Red West
AWN812V	Brewers	B155GSC	SMT	BCB610V	Rider Group	BVP819V	Badgerline
AWN813V	SWT	B156GSC	SMT	BCB611V	Rider Group	BVP820V	Badgerline
AWN815V	SWT	B157GSC	SMT	BCB612V	Rider Group	BVR52T	Rider Group
AYG848S	Eastern National	B158GSC	SMT	BCD824L	People's Provincial	BVR53T	Rider Group
AYG850S	Eastern National	B159KSC	Lowland	BCL216T	Eastern Counties	BVR55T	Rider Group
AYR299T	People's Provincial	B160KSC	Lowland	BCL217T	Eastern Counties	BVR65T	Rider Group
AYR331T	People's Provincial	B160WRN	Leicester	BDL65T	People's Provincial	BVR67T	Rider Group
AYR341T	People's Provincial	B161KSC	SMT	BEP963V	Eastern National	BVR69T	Rider Group
AYR344T	People's Provincial	B162KSC	SMT	BEP966V	Western National	BVR70T	Rider Group
B38AAF	Western National	B163KSC	SMT	BEP968V	Western National	BVR71T	Rider Group
B41AAF	Western National	B165WRN	Leicester	BEP971V	Brewers	BVR85T	Rider Group
B43AAF	Western National	B169KSC	SMT	BEP972V	Brewers	BVR87T	Rider Group
B79MJF	Leicester	B170KSC	SMT	BEP974V	Brewers	BVR92T	Rider Group
B80MJF	Leicester	B171KSC	SMT	BEP975V	SWT	BVR97T	Rider Group
B81MJF	Leicester	B172KSC	SMT	BEP976V	SWT	BYW415V	People's Provincial
B82MJF	Leicester	B173KSC	SMT	BEP977V	Brewers	C100HGL	Western National
B83MJF	Leicester	B181BLG	PMT	BEP978V	SWT	C101HGL	Western National
B84MRY	Leicester	B182BLG	PMT	BEP980V	SWT	C102HGL	Western National
B85MRY	Leicester	B188BLG	PMT	BEP981V	Brewers	C102KDS	Midland Bluebird
B86MRY	Leicester	B194BAF	Western National	BEP982V	Brewers	C103HGL	Western National
B88PKS	Midland Bluebird	B195BAF	Western National	BEP984V	SWT	C103KDS	Kirkpatrick
B93PKS	Midland Bluebird	B195BLG	PMT	BEP985V	Brewers	C104HGL	Western National
B94PKS	Midland Bluebird	B196BAF	Western National	BEP986V	Brewers	C105HGL	Western National
B95PKS	Midland Bluebird	B197BAF	Western National	BGG252S	SMT	C106HGL	Western National
B96PKS	Midland Bluebird	B199DTU	PMT	BGG253S	SMT	C107HGL	Eastern National
B98PKS	Midland Bluebird	B200DTU	PMT	BGG254S	SMT	C108HGL	Western National
B99PKS	Midland Bluebird	B201DTU	PMT	BGG255S	SMT	C108SFP	PMT
B100PKS	Midland Bluebird	B202DTU	PMT	BGG256S	SMT	C109HGL	Western National
B101PKS	Midland Bluebird	B221WEU	Brewers	BGG257S	SMT	C120VBF	PMT
B102JAB	Midland Red West	B232AFV	PMT	BGG258S	SMT	C121VRE	PMT
B102PKS	Midland Bluebird	B267KPF	Lowland	BGG259S	SMT	C122VRE	PMT
B103JAB	Midland Red West	B336BGL	Thamesway	BGG260S	SMT	C123VRE	PMT
B103PKS	Midland Bluebird	B337BGL	Thamesway	BLS423Y	Leicester	C124LHS	PMT
B104JAB	Midland Red West	B341RLS	SMT	BLS432Y	Leicester	C124VRE	PMT
B104PKS	Midland Bluebird	B448WTC	Badgerline	BLS437Y	Midland Bluebird	C125VRE	PMT
B105JAB	Midland Red West	B476YEU	Western National	BLS443Y	Leicester	C126VRE	PMT
B105PKS	Midland Bluebird	B501RWY	Rider Group	BLS446Y	Midland Bluebird	C127VRE	PMT
B106JAB	Midland Red West	B502RWY	Rider Group	BNO664T	Eastern National	C128VRE	PMT
B106PKS	Midland Bluebird	B503RWY	Rider Group	BNO666T	Thamesway	C130HJN	Eastern National
B107JAB	Midland Red West	B504RWY	Rider Group	BNO668T	Thamesway	C130VRE	PMT
B112MSO	Grampian	B505RWY	Rider Group	BNO671T	Thamesway	C131VRE	PMT
B113MSO	Grampian	B506RWY	Rider Group	BNO672T	Thamesway	C132VRE	PMT
B114MSO	Grampian	B518UWW	Rider Group	BNO674T	Thamesway	C133VRE	PMT
B115MSO	Grampian	B519UWW	Rider Group	BNO675T	Eastern National	C134VRE	PMT
B116MSO	Grampian	B520UWW	Rider Group	BNO676T	Thamesway	C135VRE	PMT
B117MSO	Grampian	B521UWW	Rider Group	BNO677T	Thamesway	C136VRE	PMT
B117OBF	PMT	B561RWY	Rider Group	BNO679T	Thamesway	C137VRE	PMT
B118MSO	Grampian	B562RWY	Rider Group	BNO680T	Eastern National	C138VRE	PMT

Reg	Operator	Reg	Operator	Reg	Operator	Reg	Operator
C139VRE	PMT	C304PNP	Midland Red West	C385RUY	Midland Red West	C674ECV	Western National
C140VRE	PMT	C305AHP	Durbin Coaches	C386RUY	Midland Red West	C675ECV	Western National
C141VRE	PMT	C305PNP	Midland Red West	C387RUY	Midland Red West	C676ECV	Western National
C142VRE	PMT	C306PNP	Midland Red West	C388RUY	Midland Red West	C677ECV	Western National
C143VRE	PMT	C307PNP	Midland Red West	C389RUY	Midland Red West	C678ECV	Eastern National
C144VRE	PMT	C308PNP	Midland Red West	C390RUY	Midland Red West	C679ECV	Western National
C145WRE	PMT	C309PNP	Midland Red West	C391RUY	Midland Red West	C680ECV	Badgerline
C146KBT	Rider Group	C310PNP	Midland Red West	C392RUY	Midland Red West	C681ECV	Badgerline
C146WRE	PMT	C311PNP	Midland Red West	C393RUY	Midland Red West	C682ECV	Western National
C147KBT	Rider Group	C312KTH	SWT	C394RUY	Midland Red West	C683ECV	Western National
C147WRE	PMT	C312PNP	Midland Red West	C395RUY	Midland Red West	C683LGE	PMT
C148KBT	Rider Group	C313PNP	Midland Red West	C396RUY	Midland Red West	C684ECV	Eastern National
C148WRE	PMT	C314PNP	Midland Red West	C397RUY	Midland Red West	C684LGE	PMT
C149KBT	Rider Group	C315PNP	Midland Red West	C398RUY	Midland Red West	C685ECV	Eastern National
C149WRE	PMT	C316PNP	Midland Red West	C399RUY	Midland Red West	C686ECV	Western National
C150KBT	Rider Group	C317PNP	Midland Red West	C400RUY	Midland Red West	C687ECV	Eastern National
C150WRE	PMT	C318PNP	Midland Red West	C401RUY	Midland Red West	C688ECV	Eastern National
C151WRE	PMT	C319PNP	Midland Red West	C402RUY	Midland Red West	C689ECV	Western National
C159DWT	Western National	C320PNP	Midland Red West	C403RUY	Midland Red West	C690ECV	Western National
C174VSF	SMT	C321PNP	Midland Red West	C404RUY	Midland Red West	C691ECV	Western National
C175VSF	SMT	C322PNP	Midland Red West	C407HJN	Eastern National	C693ECV	Western National
C176VSF	SMT	C323PNP	Midland Red West	C408HJN	Eastern National	C694ECV	Western National
C177VSF	SMT	C324PNP	Midland Red West	C409HJN	Thamesway	C695ECV	Eastern National
C178VSF	SMT	C325PNP	Midland Red West	C410HJN	Eastern National	C696ECV	Thamesway
C179VSF	SMT	C326PNP	Midland Red West	C412HJN	Eastern National	C697ECV	Eastern National
C180VSF	SMT	C327PNP	Midland Red West	C413HJN	Eastern National	C698ECV	Eastern National
C181VSF	SMT	C328PNP	Midland Red West	C414HJN	Eastern National	C699ECV	Thamesway
C182VSF	SMT	C329PNP	Midland Red West	C415HJN	Eastern National	C700ECV	Eastern National
C183VSF	SMT	C330PNP	Midland Red West	C416HJN	Eastern National	C700USC	Lowland
C201HJN	Thamesway	C331PNP	Midland Red West	C417HJN	Eastern National	C706JMB	PMT
C201HTH	Brewers	C332PNP	Midland Red West	C418HJN	Eastern National	C709JMB	PMT
C201PCD	Western National	C333PNP	Midland Red West	C419HJN	Eastern National	C710JMB	PMT
C202HJN	Eastern National	C334PNP	Midland Red West	C421HJN	Eastern National	C711BEX	Eastern Counties
C202HTH	Brewers	C335PNP	Midland Red West	C473BHY	City Line	C711JMB	PMT
C202PCD	Western National	C336PNP	Midland Red West	C475BHY	Midland Red West	C712BEX	Eastern Counties
C203HJN	Eastern National	C337PNP	Midland Red West	C476BHY	Midland Red West	C712GEV	Thamesway
C203HTH	Brewers	C338PNP	Midland Red West	C477BHY	Midland Red West	C713BEX	Eastern Counties
C204HJN	Eastern National	C339PNP	Midland Red West	C478BHY	Brewers	C714BEX	Eastern Counties
C204HTH	Brewers	C340PNP	Midland Red West	C480BHY	Brewers	C715BEX	Eastern Counties
C205HJN	Eastern National	C341PNP	Midland Red West	C481BHY	Brewers	C716BEX	Eastern Counties
C205HTH	Brewers	C342PNP	Midland Red West	C482BHY	Eastern National	C717BEX	Eastern Counties
C206HJN	Eastern National	C343PNP	Midland Red West	C483BFB	Thamesway	C718BEX	Eastern Counties
C206HTH	Brewers	C344PNP	Midland Red West	C483BHY	Midland Red West	C719BEX	Eastern Counties
C207HJN	Eastern National	C345PNP	Midland Red West	C483YWY	Rider Group	C720BEX	Eastern Counties
C207HTH	Brewers	C346PNP	Midland Red West	C484BHY	Eastern National	C721BEX	Eastern Counties
C207PCD	Midland Red West	C347PNP	Midland Red West	C485BHY	Eastern National	C722BEX	Eastern Counties
C208HJN	Western National	C348PNP	Midland Red West	C485YWY	Rider Group	C723BEX	Eastern Counties
C208HTH	Brewers	C349PNP	Midland Red West	C486BHY	Eastern National	C724BEX	Eastern Counties
C208PCD	Midland Red West	C350PNP	Midland Red West	C487BFB	Badgerline	C725BEX	Eastern Counties
C209HJN	Brewers	C351PNP	Midland Red West	C487BHY	Midland Red West	C725FKE	Eastern Counties
C209HTH	Brewers	C352PNP	Midland Red West	C488BHY	Midland Red West	C726BEX	Eastern Counties
C209PCD	Midland Red West	C353PNP	Midland Red West	C489BHY	Eastern National	C726FKE	Eastern Counties
C210HJN	Eastern National	C354PNP	Midland Red West	C489HCV	Western National	C727BEX	Eastern Counties
C210HTH	Brewers	C355PNP	Midland Red West	C490BHY	Midland Red West	C727FKE	Eastern Counties
C211HTH	Brewers	C356PNP	Midland Red West	C491BHY	Midland Red West	C741BEX	Eastern Counties
C211PCD	Western National	C357PNP	Midland Red West	C491HCV	Thamesway	C742BEX	Eastern Counties
C212HJN	Eastern National	C358PNP	Midland Red West	C492BHY	Midland Red West	C743BEX	Eastern Counties
C212HTH	Brewers	C359PNP	Midland Red West	C493BHY	Eastern National	C744BEX	Eastern Counties
C212PCD	Midland Red West	C360PNP	Midland Red West	C494BHY	Eastern National	C745BEX	Eastern Counties
C213HJN	Western National	C361RUY	Midland Red West	C495BHY	Eastern National	C746BEX	Eastern Counties
C213HTH	Brewers	C362RUY	Midland Red West	C496BHY	Eastern National	C747BEX	Eastern Counties
C214HJN	Western National	C363RUY	Midland Red West	C497BHY	Midland Red West	C748BEX	Eastern Counties
C214HTH	Brewers	C364RUY	Midland Red West	C498BHY	Midland Red West	C749BEX	Eastern Counties
C215HJN	Eastern National	C365RUY	Midland Red West	C499BHY	Midland Red West	C750BEX	Eastern Counties
C215HTH	SWT	C366RUY	Midland Red West	C507KBT	Rider Group	C751BEX	Eastern Counties
C216HJN	Western National	C367RUY	Midland Red West	C508KBT	Rider Group	C752BEX	Eastern Counties
C217HJN	Eastern National	C368RUY	Midland Red West	C509KBT	Rider Group	C753BEX	Eastern Counties
C218HJN	Western National	C369RUY	Midland Red West	C510KBT	Rider Group	C754BEX	Eastern Counties
C219HJN	Western National	C370RUY	Midland Red West	C511KBT	Rider Group	C755BEX	Eastern Counties
C220HJN	Western National	C371RUY	Midland Red West	C581SHC	Midland Red West	C756BEX	Eastern Counties
C221HJN	Brewers	C372RUY	Midland Red West	C582SHC	Midland Red West	C757BEX	Eastern Counties
C222HJN	Rider Group	C373RUY	Midland Red West	C583SHC	Midland Red West	C783FRL	Western National
C223HJN	Western National	C374RUY	Midland Red West	C584SHC	Midland Red West	C784FRL	Western National
C224HJN	Western National	C375RUY	Midland Red West	C585SHC	Midland Red West	C785FRL	Western National
C229HCV	Western National	C376RUY	Midland Red West	C586SHC	Midland Red West	C786FRL	Western National
C230HCV	Eastern National	C377RUY	Midland Red West	C587SHC	Midland Red West	C787FRL	Western National
C231HCV	Eastern National	C378RUY	Midland Red West	C588SHC	Midland Red West	C788FRL	Midland Red West
C232HCV	Eastern National	C379RUY	Midland Red West	C589SHC	Midland Red West	C789FRL	Western National
C28EUH	Badgerline	C380RUY	Midland Red West	C637BEX	Eastern Counties	C790FRL	Midland Red West
C29EUH	Badgerline	C381RUY	Midland Red West	C638BEX	Eastern Counties	C791FRL	Western National
C301PNP	Midland Red West	C382RUY	Midland Red West	C652BEX	Eastern Counties	C792FRL	Western National
C302PNP	Midland Red West	C383RUY	Midland Red West	C672ECV	Western National	C793FRL	Western National
C303PNP	Midland Red West	C384RUY	Midland Red West	C673ECV	Western National	C794FRL	Western National

Reg	Operator	Reg	Operator	Reg	Operator	Reg	Operator
C795FRL	Western National	CJO470R	Eastern Counties	D100GHY	Badgerline	D225PPU	Thamesway
C796FRL	Western National	CJO471R	Eastern Counties	D101GHY	Badgerline	D226LCY	SWT
C797FRL	Western National	CJO472R	Eastern Counties	D101XNV	Northampton	D226UHC	Midland Bluebird
C798FRL	Western National	CNH48T	Northampton	D102GHY	Badgerline	D227LCY	SWT
C799FRL	Western National	CNH49T	Northampton	D102XNV	Northampton	D227UHC	Midland Bluebird
C800FRL	Western National	CNH50T	Northampton	D103GHY	Badgerline	D228LCY	SWT
C801FRL	Western National	CNH52T	Northampton	D104GHY	Badgerline	D229LCY	SWT
C802FRL	Western National	CNH53T	Northampton	D105GHY	Badgerline	D229UHC	Midland Bluebird
C805SDY	Kirkpatrick	CNH54T	Northampton	D106GHY	Badgerline	D230LCY	SWT
C812SDY	Midland Bluebird	CNH55T	Northampton	D107GHY	Badgerline	D230PPU	Thamesway
C817SDY	Midland Bluebird	CNH57T	Northampton	D108ELS	Midland Bluebird	D231PPU	Thamesway
C821SDY	Midland Bluebird	CNH58T	Northampton	D108GHY	Badgerline	D232LCY	SWT
C892BEX	Eastern Counties	CPT823S	Western National	D109ELS	Midland Bluebird	D232PPU	Thamesway
C894BEX	Eastern Counties	CRG325C	Grampian	D109GHY	Badgerline	D232UHC	Mair's
C895BEX	Eastern Counties	CSG773S	SMT	D110ELS	Midland Bluebird	D233LCY	SWT
C896BEX	Eastern Counties	CSG774S	SMT	D110GHY	Badgerline	D233PPU	Thamesway
C897BEX	Eastern Counties	CSG775S	SMT	D111ELS	Midland Bluebird	D234LCY	SWT
C898BEX	Eastern Counties	CSG776S	SMT	D111GHY	Badgerline	D234PPU	Thamesway
C901FCY	SWT	CSG777S	SMT	D112GHY	Badgerline	D235LCY	SWT
C902FCY	SWT	CSG778S	SMT	D113GHY	Badgerline	D235PPU	Thamesway
C903BEX	Eastern Counties	CSG779S	SMT	D115ELS	Midland Bluebird	D236LCY	SWT
C903FCY	SWT	CSG780S	SMT	D116ELS	Midland Bluebird	D236PPU	Thamesway
C904FCY	SWT	CSG782S	SMT	D118DRV	People's Provincial	D237LCY	SWT
C905BEX	Eastern Counties	CSG792S	Lowland	D118PGA	PMT	D237PPU	Thamesway
C905FCY	SWT	CSG794S	Lowland	D119DRV	People's Provincial	D238LCY	SWT
C906BEX	Eastern Counties	CSU244	Rider Group	D119NUS	Midland Bluebird	D238PPU	Thamesway
C906FCY	SWT	CUB21Y	Rider Group	D119PGA	PMT	D239LCY	SWT
C907FCY	SWT	CUB22Y	Rider Group	D120DRV	People's Provincial	D239PPU	Thamesway
C908BEX	Eastern Counties	CUB23Y	Rider Group	D120NUS	Midland Bluebird	D240LCY	SWT
C912BEX	Eastern Counties	CUB24Y	Rider Group	D120PGA	PMT	D240PPU	Thamesway
C913BEX	Eastern Counties	CUB25Y	Rider Group	D121DRV	People's Provincial	D241LCY	SWT
C919BEX	Eastern Counties	CUB26Y	Rider Group	D121PGA	PMT	D241PPU	Thamesway
C949GAF	Western National	CUB27Y	Rider Group	D122DRV	People's Provincial	D242LCY	SWT
C950GAF	Western National	CUB28Y	Rider Group	D122PGA	PMT	D242PPU	Thamesway
C951GAF	Western National	CUB29Y	Rider Group	D123DRV	People's Provincial	D243LCY	SWT
C952GAF	Western National	CUB30Y	Rider Group	D123PGA	PMT	D243PPU	Thamesway
C952YAH	Eastern Counties	CUB31Y	Rider Group	D124PGA	PMT	D244LCY	SWT
C953GAF	Western National	CUB32Y	Rider Group	D127DRV	People's Provincial	D244PPU	Thamesway
C954GAF	Western National	CUB33Y	Rider Group	D138NUS	City Line	D245LCY	SWT
C955GAF	Western National	CUB34Y	Rider Group	D144NDT	Durbin Coaches	D246LCY	SWT
C956GAF	Western National	CUB35Y	Rider Group	D152BEH	PMT	D247LCY	SWT
C957GAF	Thamesway	CUB36Y	Rider Group	D153BEH	PMT	D248LCY	SWT
C957YAH	Eastern Counties	CUB37Y	Rider Group	D154BEH	PMT	D249LCY	SWT
C958GAF	Western National	CUB38Y	Rider Group	D155BEH	PMT	D250LCY	SWT
C958YAH	Eastern Counties	CUB39Y	Rider Group	D156BEH	PMT	D251LCY	SWT
C959GAF	Western National	CUB40Y	Rider Group	D157BEH	PMT	D252LCY	SWT
C963GCV	Thamesway	CUB41Y	Rider Group	D157VRP	PMT	D253LCY	SWT
C964GCV	Eastern Counties	CUB42Y	Rider Group	D158BEH	PMT	D276FAS	Lowland
C965GCV	Western National	CUB43Y	Rider Group	D159BEH	PMT	D321REF	Leicester
C967GCV	Thamesway	CUB44Y	Rider Group	D159VRP	PMT	D328DKS	Lowland
C968GCV	Western National	CUB45Y	Rider Group	D160VRP	PMT	D329DKS	Lowland
C970GCV	Western National	CUB46Y	Rider Group	D162LTA	PMT	D330DKS	Lowland
C972YAH	Eastern Counties	CUB47Y	Rider Group	D162VRP	PMT	D349ESC	SMT
C974GCV	Brewers	CUB48Y	Rider Group	D165VRP	PMT	D350ESC	SMT
C975GCV	Brewers	CUB51Y	Rider Group	D176VRP	PMT	D351ESC	SMT
C975YAH	Eastern Counties	CUB52Y	Rider Group	D178VRP	PMT	D401ASF	Eastern Counties
C976GCV	Brewers	CUB53Y	Rider Group	D179VRP	PMT	D402ASF	Lowland
C977GCV	Brewers	CUB54Y	Rider Group	D180BEH	PMT	D403ASF	Lowland
C978GCV	Thamesway	CUB55Y	Rider Group	D180VRP	PMT	D404ASF	SMT
C979GCV	Badgerline	CUB65Y	Rider Group	D182BEH	PMT	D405ASF	SMT
C980GCV	Badgerline	CUB521Y	Rider Group	D183BEH	PMT	D406ASF	SMT
C981GCV	Badgerline	CUB522Y	Rider Group	D184BEH	PMT	D407ASF	Eastern Counties
C982GCV	Badgerline	CUB523Y	Rider Group	D184ESC	SMT	D408ASF	Lowland
C982YAH	Eastern Counties	CUB525Y	Rider Group	D184VRP	PMT	D409ASF	SMT
C983GCV	Badgerline	CUB527Y	Rider Group	D185BEH	PMT	D410ASF	Lowland
C984GCV	Badgerline	CUB530Y	Rider Group	D185ESC	SMT	D411ASF	SMT
C984YAH	Eastern Counties	CUB531Y	Rider Group	D185VRP	PMT	D412ASF	SMT
C985GCV	Badgerline	CUB532Y	Rider Group	D186BEH	PMT	D413ASF	SMT
C985HOX	Midland Red West	CUB534Y	Rider Group	D186ESC	SMT	D414ASF	SMT
C986GCV	Badgerline	CUB535Y	Rider Group	D186VRP	PMT	D415ASF	Eastern Counties
C986HOX	Midland Red West	CUB536Y	Rider Group	D187BEH	PMT	D416ASF	SMT
C987GCV	Badgerline	CUB537Y	Rider Group	D187VRP	PMT	D417ASF	Eastern Counties
C987HOX	Midland Red West	CWU137T	Rider Group	D188BEH	PMT	D418ASF	SMT
C988GCV	Badgerline	CWU144T	Rider Group	D189BEH	PMT	D419ASF	SMT
C989GCV	Badgerline	CWU148T	Rider Group	D216LCY	Brewers	D420ASF	SMT
C98HGL	Western National	CWU149T	Rider Group	D217LCY	Brewers	D421ASF	SMT
C990GCV	Eastern National	CWU152T	Rider Group	D219LCY	Brewers	D422ASF	SMT
C99HGL	Western National	CWU153T	Rider Group	D220LCY	SWT	D423ASF	SMT
CCL774T	Eastern Counties	CWU155T	Rider Group	D221LCY	SWT	D424ASF	SMT
CCL775T	Eastern Counties	CWU156T	Rider Group	D222LCY	SWT	D425ASF	Eastern Counties
CCL776T	Eastern Counties	D33XSS	Grampian	D223LCY	SWT	D426ASF	SMT
CCL778T	Eastern Counties	D44OYA	Western National	D224LCY	SWT	D427ASF	SMT
CFS155W	Midland Bluebird	D83KRL	Western National	D225LCY	SWT	D428ASF	SMT

Reg	Operator	Reg	Operator	Reg	Operator	Reg	Operator
D429ASF	SMT	D700GHY	Badgerline	DSA241T	Grampian	E282TTH	SWT
D430ASF	Eastern Counties	D701GHY	Badgerline	DSA242T	Grampian	E283UCY	Brewers
D451ERE	PMT	D702GHY	Badgerline	DSA243T	Grampian	E284UCY	Brewers
D452ERE	PMT	D703GHY	Badgerline	DSA244T	Grampian	E285UCY	Brewers
D453ERE	PMT	D704GHY	Badgerline	DSA245T	Grampian	E286UCY	Brewers
D454ERE	PMT	D705GHY	Badgerline	DSA246T	Grampian	E287UCY	SWT
D455ERE	PMT	D706GHY	Badgerline	DSA247T	Grampian	E288VEP	Brewers
D456ERE	PMT	D707GHY	Badgerline	DSA248T	Grampian	E289VEP	SWT
D457ERE	PMT	D708GHY	Badgerline	DSA249T	Grampian	E290VEP	SWT
D458ERE	PMT	D709GHY	Badgerline	DSA250T	Grampian	E291VEP	SWT
D459CKV	Lowland	D710GHY	Badgerline	DSA251T	Grampian	E292VEP	SWT
D459ERE	PMT	D711CKS	Lowland	DSA252T	Grampian	E293VEP	SWT
D500FAE	SWT	D711GHY	Badgerline	DSA253T	Grampian	E294VEP	SWT
D500GHY	Badgerline	D712CKS	Lowland	DSA254T	Grampian	E295VEP	SWT
D501GHY	Badgerline	D713CSH	Lowland	DSA256T	Grampian	E296VEP	SWT
D502GHY	Badgerline	D714CSH	Lowland	DSA257T	Grampian	E297VEP	SWT
D503GHY	Badgerline	D727GDE	Rider Group	DSD958V	Lowland	E298VEP	SWT
D504FAE	Thamesway	D750DSH	Lowland	DSD965V	Lowland	E299VEP	SWT
D505FAE	Thamesway	D751DSH	Lowland	DSD967V	Lowland	E300VEP	SWT
D506FAE	Thamesway	D752DSH	Lowland	DSD968V	Lowland	E301VEP	SWT
D507FAE	Thamesway	D753DSH	Lowland	DSX400S	Lowland	E302VEP	SWT
D508FAE	Durbin Coaches	D754DSH	Lowland	DWU296T	Rider Group	E303VEP	SWT
D509FAE	Durbin Coaches	D755DSH	Lowland	DWU298T	Eastern National	E304VEP	SWT
D510PPU	Eastern National	D756RWC	Wessex	E39KRE	PMT	E305VEP	SWT
D511FAE	Thamesway	D758LEX	Eastern Counties	E41JRF	PMT	E306VEP	SWT
D511PPU	Eastern National	D759LEX	Eastern Counties	E43JRF	PMT	E326SWY	Rider Group
D512HUB	Rider Group	D763KWT	Midland Red West	E44JRF	PMT	E330SWY	Rider Group
D512PPU	Eastern National	D764KWT	Thamesway	E60MMT	King's of Dunblain	E331SWY	Rider Group
D513HUB	Rider Group	D779NUD	Eastern Counties	E87HNR	Leicester	E336SWY	Rider Group
D514FAE	Brewers	D783SGB	Rider Group	E88HNR	Leicester	E337SWY	Rider Group
D514HUB	Rider Group	D793KWR	Rider Group	E89HNR	Leicester	E339SWY	Rider Group
D515FAE	SWT	D796KWR	Rider Group	E90HNR	Leicester	E342NFA	PMT
D515HUB	Rider Group	D799KWR	Rider Group	E91HNR	Leicester	E342SWY	Rider Group
D516HUB	Rider Group	D862LWR	Rider Group	E92HNR	Leicester	E345SWY	Rider Group
D517FAE	Brewers	D865LWR	Rider Group	E93HNR	Leicester	E350AMR	Badgerline
D520FAE	Brewers	D868LWR	Rider Group	E94HNR	Leicester	E384XCA	PMT
D526FAE	Brewers	D869LWR	Rider Group	E95HNR	Leicester	E385CNE	Rider Group
D527FAE	Brewers	D870LWR	Rider Group	E96HNR	Leicester	E400HWC	Thamesway
D529FAE	Rider Group	D871LWR	Rider Group	E97HNR	Leicester	E401HWC	Eastern National
D530FAE	Rider Group	D874LWR	Rider Group	E98HNR	Leicester	E406HAB	Midland Red West
D531FAE	Rider Group	D890MDB	Leicester	E99HNR	Leicester	E406RWR	Rider Group
D532FAE	Rider Group	D901CSH	Lowland	E106JNH	Grampian	E407HAB	Midland Red West
D532RCK	Mair's	D901GEU	Badgerline	E106LVT	PMT	E408HAB	Midland Red West
D533FAE	SWT	D902CSH	Lowland	E108JNH	Grampian	E409HAB	Midland Red West
D534KGL	Eastern National	D902HOU	Badgerline	E110JNH	Grampian	E410HAB	Midland Red West
D535FAE	Rider Group	D903HOU	Badgerline	E111NNV	Northampton	E411HAB	Midland Red West
D536FAE	Rider Group	D904HOU	Badgerline	E122DRS	Grampian	E412KUY	Midland Red West
D537FAE	Rider Group	D905HOU	Badgerline	E123DRS	Grampian	E413KUY	Midland Red West
D539FAE	Rider Group	D906HOU	Badgerline	E124DRS	Grampian	E414KUY	Midland Red West
D540FAE	Rider Group	D909HOU	Badgerline	E125DRS	Grampian	E415KUY	Midland Red West
D541FAE	Rider Group	D951VSS	Kirkpatrick	E126DRS	Grampian	E416KUY	Midland Red West
D542FAE	SWT	DAR118T	Thamesway	E127DRS	Grampian	E417KUY	Midland Red West
D543FAE	SWT	DAR119T	Thamesway	E128DRS	Grampian	E418KUY	Midland Red West
D544FAE	SWT	DAR120T	Thamesway	E129DRS	Grampian	E419KUY	Midland Red West
D545FAE	SWT	DAR121T	Eastern National	E130DRS	Grampian	E420KUY	Midland Red West
D546FAE	SWT	DAR125T	Thamesway	E131DRS	Grampian	E421KUY	Midland Red West
D547FAE	SWT	DAR128T	Thamesway	E166CNC	Rider Group	E422KUY	Midland Red West
D548FAE	PMT	DAR129T	Thamesway	E187HSF	SMT	E423KUY	Midland Red West
D549FAE	PMT	DAR130T	Thamesway	E188HSF	SMT	E424KUY	Midland Red West
D550FAE	PMT	DAR131T	Thamesway	E189HSF	SMT	E425KUY	Midland Red West
D551FAE	PMT	DAR133T	Thamesway	E190HSF	SMT	E426KUY	Midland Red West
D552FAE	SWT	DAR134T	Thamesway	E200BOD	Western National	E427KUY	Midland Red West
D554FAE	Rider Group	DCA526X	PMT	E201BOD	Western National	E428KUY	Midland Red West
D555FAE	Rider Group	DEX226T	Eastern Counties	E201PWY	Rider Group	E429KUY	Midland Red West
D556FAE	Rider Group	DEX229T	Eastern Counties	E202BOD	Western National	E430KUY	Midland Red West
D557FAE	SWT	DEX230T	Eastern Counties	E202PWY	Rider Group	E431JSG	SMT
D558FAE	SWT	DFB672W	Durbin Coaches	E203BOD	Western National	E431KUY	Midland Red West
D559FAE	SWT	DFS806S	Lowland	E203PWY	Rider Group	E432JSG	SMT
D560FAE	Rider Group	DFS807S	Lowland	E204BOD	Western National	E432KUY	Midland Red West
D561FAE	Rider Group	DHW349W	Badgerline	E204PWY	Rider Group	E433JSG	SMT
D562FAE	Rider Group	DHW351W	Badgerline	E205BOD	Western National	E433KUY	Lowland
D563FAE	Rider Group	DLS351V	Midland Bluebird	E206BOD	SWT	E434JSG	Lowland
D564FAE	Rider Group	DLS352V	Midland Bluebird	E208BOD	Badgerline	E434KUY	Lowland
D565FAE	SWT	DLS353V	Midland Bluebird	E209BOD	Badgerline	E435JSG	Lowland
D566FAE	Rider Group	DLS355V	Midland Bluebird	E210BOD	Badgerline	E435KUY	Midland Red West
D567FAE	Rider Group	DLS356V	Midland Bluebird	E211BOD	Badgerline	E436JSG	SMT
D568FAE	Rider Group	DMS26V	Midland Bluebird	E212BOD	Badgerline	E436KUY	Midland Red West
D569FAE	Rider Group	DNG236T	Eastern Counties	E215BTA	Badgerline	E437JSG	SMT
D591MVR	King's of Dunblain	DPW781T	Eastern Counties	E216BTA	Badgerline	E437KUY	Midland Red West
D599MVR	King's of Dunblain	DPW782T	Eastern Counties	E217BTA	Badgerline	E438JSG	SMT
D600GHY	Badgerline	DSA238T	Grampian	E279TTH	SWT	E438KUY	Midland Red West
D601GHY	Badgerline	DSA239T	Grampian	E280TTH	SWT	E439JSG	SMT
D617YCX	Western National	DSA240T	Grampian	E281TTH	SWT	E439KUY	Midland Red West

Reg	Operator	Reg	Operator	Reg	Operator	Reg	Operator
E440JSG	Lowland	E807HBF	PMT	ERP558T	Brewers	F148MBC	Leicester
E442JSG	SMT	E807MOU	Western National	ESC843S	Lowland	F149MBC	Leicester
E443JSG	SMT	E808HBF	PMT	ESC846S	Lowland	F150MBC	Leicester
E444JSG	SMT	E808MOU	Western National	ESC848S	Lowland	F151MBC	Leicester
E446JSG	SMT	E809HBF	PMT	ESC849S	Lowland	F151XYG	Rider Group
E447JSG	SMT	E809MOU	Badgerline	ESK955	Grampian	F152MBC	Leicester
E448JSG	SMT	E810HBF	PMT	ESK956	Grampian	F152XYG	Rider Group
E449JSG	SMT	E810MOU	Badgerline	ESK957	Grampian	F153XYG	Rider Group
E450JSG	SMT	E811HBF	PMT	ESK958	Midland Bluebird	F154XYG	Rider Group
E451JSG	SMT	E811MOU	Badgerline	ESU980	Durbin Coaches	F155XYG	Rider Group
E452JSG	SMT	E812HBF	PMT	ESX257	Lowland	F156XYG	Rider Group
E453JSG	SMT	E812MOU	Western National	ETH68V	Western National	F157XYG	Rider Group
E454JSG	SMT	E813HBF	PMT	EWN992W	SWT	F158XYG	Rider Group
E455JSG	SMT	E813MOU	Badgerline	EWN994W	SWT	F159XYG	Rider Group
E456JSG	SMT	E814HBF	PMT	EWN995W	SWT	F160XYG	Rider Group
E457JSG	SMT	E814MOU	Badgerline	EWR165T	Rider Group	F161XYG	Rider Group
E458JSG	SMT	E815HBF	PMT	EWR651Y	Eastern National	F162XYG	Rider Group
E459JSG	SMT	E815MOU	Badgerline	EWR652Y	Eastern National	F163XYG	Rider Group
E460JSG	SMT	E816HBF	PMT	EWR653Y	Eastern National	F164XYG	Rider Group
E461JSG	SMT	E816MOU	Badgerline	EWR654Y	Rider Group	F165XYG	Rider Group
E462JSG	SMT	E817HBF	PMT	EWR655Y	Rider Group	F166ONT	PMT
E463JSG	SMT	E817MOU	Badgerline	EWR656Y	Rider Group	F166XYG	Rider Group
E464JSG	SMT	E818HBF	PMT	EWR657Y	Rider Group	F167XYG	Rider Group
E466JSG	SMT	E818MOU	Badgerline	EWS739W	Badgerline	F168XYG	Rider Group
E467JSG	Kirkpatrick	E819HBF	PMT	EWS741W	Badgerline	F169XYG	Rider Group
E467TYG	Rider Group	E819MOU	Badgerline	EWS742W	Badgerline	F170XYG	Rider Group
E468JSG	SMT	E820HBF	PMT	EWS744W	Badgerline	F171XYG	Rider Group
E469JSG	SMT	E820MOU	Western National	EWS745W	Badgerline	F172XYG	Rider Group
E470JSG	SMT	E821HBF	PMT	EWS747W	Western National	F173XYG	Rider Group
E470MVT	PMT	E821MOU	Badgerline	EWS749W	Badgerline	F174XYG	Rider Group
E471MVT	PMT	E822HBF	PMT	EWS750W	Badgerline	F175XYG	Rider Group
E471TYG	Rider Group	E822MOU	Badgerline	EWS752W	Badgerline	F217OFB	PMT
E526NEH	PMT	E823MOU	Badgerline	EWS753W	Badgerline	F229FSU	Rider Group
E527JRE	PMT	E824HBF	PMT	EWS754W	Badgerline	F245MVW	Thamesway
E528JRE	PMT	E825HBF	PMT	EWV665	Western National	F246MVW	Thamesway
E570NFB	City Line	E826HBF	PMT	EWW945Y	Rider Group	F247NJN	Thamesway
E571NFB	City Line	E831ETY	PMT	EWW946Y	Eastern National	F248NJN	Thamesway
E572NFB	City Line	E851PEX	Eastern Counties	EWY77Y	Rider Group	F249NJN	Thamesway
E573NFB	City Line	E852PEX	Eastern Counties	EWY78Y	PMT	F250NJN	Thamesway
E574NFB	City Line	E853PEX	Eastern Counties	EWY79Y	PMT	F251NJN	Thamesway
E575NFB	City Line	E854PEX	Eastern Counties	EWY81Y	Rider Group	F252NJN	Thamesway
E576NFB	City Line	E855PEX	Eastern Counties	F21URL	Western National	F253RHK	Thamesway
E577NFB	City Line	E856PEX	Eastern Counties	F22URL	Western National	F254RHK	Thamesway
E580OOU	City Line	E922KEU	Badgerline	F83XBD	Northampton	F255RHK	Thamesway
E581OOU	City Line	E923KEU	Badgerline	F84XBD	Northampton	F256RHK	Thamesway
E582OOU	City Line	E924KEU	Badgerline	F85XBD	Northampton	F257RHK	Thamesway
E583OOU	City Line	E925KEU	Badgerline	F86DVV	Northampton	F258RHK	Thamesway
E584OOU	City Line	E927KEU	Badgerline	F87DVV	Northampton	F259RHK	Thamesway
E585OOU	City Line	E928KEU	Badgerline	F88CWG	PMT	F260RHK	Thamesway
E586OOU	City Line	E929KEU	Badgerline	F88DVV	Northampton	F300GNS	Rider Group
E587OOU	City Line	E930KEU	Badgerline	F95CWG	PMT	F307AWN	SWT
E588OOU	City Line	E931KEU	Badgerline	F99CEP	SWT	F308AWN	SWT
E589OOU	City Line	E932KEU	Badgerline	F100CEP	SWT	F309AWN	SWT
E675UNE	Thamesway	E934KEU	Badgerline	F100UEH	PMT	F310AWN	SWT
E677UNE	Thamesway	E935KEU	Badgerline	F101AVG	Eastern Counties	F310REH	PMT
E683XVU	Western National	E936KEU	Badgerline	F102AVG	Eastern Counties	F311AWN	SWT
E694UND	Durbin Coaches	E937KEU	Badgerline	F103AVG	Eastern Counties	F311REH	PMT
E701TNG	Eastern Counties	E939KEU	Badgerline	F104AVG	Eastern Counties	F312AWN	SWT
E702TNG	Eastern Counties	E940KEU	Badgerline	F105AVG	Eastern Counties	F312REH	PMT
E756GSH	Lowland	E942LAE	Badgerline	F106CWG	PMT	F313AWN	SWT
E760HBF	PMT	E943LAE	Badgerline	F107CWG	PMT	F313REH	PMT
E761HBF	PMT	E944LAE	Wessex	F108CWG	PMT	F314AWN	SWT
E762HBF	PMT	E945LAE	Badgerline	F109CWG	PMT	F314REH	PMT
E764HBF	PMT	E952LAE	Badgerline	F110CWG	PMT	F315AWN	SWT
E765HBF	PMT	E976MFB	City Line	F128SBP	People's Provincial	F315REH	PMT
E766HBF	PMT	EAH890Y	Midland Red West	F129SBP	People's Provincial	F316AWN	SWT
E767HBF	PMT	ECY988V	Brewers	F130SBP	People's Provincial	F316REH	PMT
E768HBF	PMT	ECY989V	SWT	F131SBP	People's Provincial	F317AWN	SWT
E769HBF	PMT	ECY990V	SWT	F132SBP	People's Provincial	F317REH	PMT
E791CCA	PMT	ECY991V	SWT	F133SBP	People's Provincial	F318AWN	Brewers
E800MOU	Badgerline	EEL893V	People's Provincial	F134DEP	SWT	F318AWN	SWT
E801HBF	PMT	EHE234V	Eastern Counties	F134TCR	People's Provincial	F319AWN	SWT
E801MOU	Badgerline	EJO490V	PMT	F135DEP	SWT	F320AWN	SWT
E802HBF	PMT	EJO491V	PMT	F135TCR	People's Provincial	F321AWN	SWT
E802MOU	Badgerline	EJO492V	PMT	F136TCR	People's Provincial	F322AWN	SWT
E803HBF	PMT	EMB358S	PMT	F140MBC	Leicester	F323DCY	SWT
E803MOU	Badgerline	EMS360V	Midland Bluebird	F141MBC	Leicester	F324DCY	SWT
E804HBF	PMT	EMS362V	Midland Bluebird	F142MBC	Leicester	F325DCY	SWT
E804MOU	Badgerline	EMS363V	Midland Bluebird	F143MBC	Leicester	F326DCY	SWT
E805HBF	PMT	EMS364V	Midland Bluebird	F144MBC	Leicester	F326WCS	Mair's
E805MOU	Badgerline	EMS366V	Midland Bluebird	F145MBC	Leicester	F327DCY	SWT
E806HBF	PMT	EPH227V	Rider Group	F146MBC	Leicester	F328FCY	SWT
E806MOU	Badgerline	ERF24Y	PMT	F147MBC	Leicester	F329FCY	SWT
						F330FCY	SWT

Reg	Operator	Reg	Operator	Reg	Operator	Reg	Operator
F331FCY	SWT	F599PWS	City Line	F702MBC	Eastern Counties	FUT185V	Leicester
F332FCY	SWT	F599XWY	Rider Group	F703MBC	Eastern Counties	FUT186V	Leicester
F333FCY	SWT	F600PWS	City Line	F704MBC	Eastern Counties	FUT187V	Leicester
F334FCY	SWT	F600RTC	City Line	F705MBC	Eastern Counties	FUT240V	Leicester
F335FCY	SWT	F600XWY	Rider Group	F706MBC	Eastern Counties	FUT241V	Leicester
F336FCY	SWT	F601AWN	Brewers	F707MBC	Eastern Counties	FUT244V	Leicester
F337FCY	SWT	F601PWS	City Line	F708MBC	Eastern Counties	FUT245V	Leicester
F338FCY	SWT	F601RTC	City Line	F709NJF	Eastern Counties	FUT247V	Leicester
F339FCY	SWT	F601XWY	Rider Group	F710NJF	Eastern Counties	FUT250V	Leicester
F340FCY	SWT	F602AWN	Brewers	F713OFH	PMT	FWA473V	Rider Group
F341FCY	SWT	F602PWS	City Line	F715PFP	Eastern Counties	FXI8653	PMT
F342FCY	SWT	F602RTC	City Line	F716PFP	Eastern Counties	G76RGG	Rider Group
F343FCY	SWT	F602XWY	Rider Group	F716SML	King's of Dunblain	G101EVT	PMT
F361YTJ	PMT	F603AWN	Brewers	F718PFP	Leicester	G101HNP	Midland Red West
F362YTJ	PMT	F603PWS	City Line	F719PFP	Eastern Counties	G101RSH	Lowland
F363YTJ	PMT	F603RTC	City Line	F720PFP	Leicester	G102HNP	Midland Red West
F364YTJ	PMT	F603XWY	Rider Group	F721PFP	Leicester	G103HNP	Midland Red West
F402LTW	Eastern National	F604AWN	Brewers	F722PFP	Eastern Counties	G104HNP	Midland Red West
F403LTW	Eastern National	F604PWS	City Line	F723PFP	Leicester	G105HNP	Midland Red West
F404LTW	Thamesway	F604RTC	City Line	F724PFP	Leicester	G106HNP	Midland Red West
F405LTW	Thamesway	F604XWY	Rider Group	F725PFP	Leicester	G107HNP	Midland Red West
F406LTW	Thamesway	F605AWN	Brewers	F726PFP	Leicester	G108HNP	Midland Red West
F407LTW	Eastern National	F605PWS	City Line	F767XNH	Western National	G109HNP	Midland Red West
F408LTW	Eastern National	F605RTC	City Line	F771GNA	Thamesway	G110HNP	Midland Red West
F409LTW	Thamesway	F605XWY	Rider Group	F800RHK	Thamesway	G111HNP	Midland Red West
F410MNO	Thamesway	F606AWN	Brewers	F801RHK	Thamesway	G112ENV	Northampton
F411MNO	Thamesway	F606PWS	City Line	F802RHK	Thamesway	G112HNP	Midland Red West
F412MNO	Thamesway	F606RTC	City Line	F803RHK	Thamesway	G113ENV	Northampton
F413MNO	Eastern National	F607AWN	SWT	F804RHK	Thamesway	G113HNP	Midland Red West
F414MNO	Eastern National	F607PWS	City Line	F850TCW	Badgerline	G114ENV	Northampton
F415MWC	Eastern National	F607RTC	City Line	F907DHB	Brewers	G114HNP	Midland Red West
F416MWC	Eastern National	F608AWN	Brewers	F999UGL	Western National	G115HNP	Midland Red West
F417MWC	Thamesway	F608PWS	City Line	FBC1C	Durbin Coaches	G116HNP	Midland Red West
F418MWC	Thamesway	F608RTC	City Line	FDV794V	Western National	G117HNP	Midland Red West
F419MWC	Thamesway	F608WBV	PMT	FDV798V	Western National	G118HNP	Midland Red West
F420MJN	Thamesway	F609PWS	City Line	FDV800V	Western National	G119HNP	Midland Red West
F421MJN	Thamesway	F609RTC	City Line	FDV801V	Western National	G120HNP	Midland Red West
F422MJN	Thamesway	F610PWS	City Line	FDV802V	Western National	G121HNP	Midland Red West
F423MJN	Thamesway	F610RTC	City Line	FDV805V	Western National	G122HNP	Midland Red West
F424MJN	Thamesway	F611PWS	City Line	FDV806V	Western National	G123HNP	Midland Red West
F425MJN	Eastern National	F611RTC	City Line	FDV807V	Western National	G124HNP	Midland Red West
F426MJN	Eastern National	F612PWS	City Line	FDV808V	Western National	G125HNP	Midland Red West
F427MJN	Eastern National	F612RTC	City Line	FDV814V	Western National	G126HNP	Midland Red West
F428MJN	Eastern National	F613PWS	City Line	FDV815V	Western National	G127HNP	Midland Red West
F429MJN	Eastern National	F613RTC	City Line	FDZ980	Western National	G128HNP	Midland Red West
F444DUG	Western National	F613XWY	Thamesway	FDZ984	Thamesway	G129HNP	Midland Red West
F445DUG	Western National	F614PWS	City Line	FEH1Y	Midland Red West	G130HNP	Midland Red West
F446DUG	Western National	F614RTC	City Line	FFK312	Leicester	G131HNP	Midland Red West
F452YHF	PMT	F614XWY	Thamesway	FFS6X	Lowland	G132HNP	Midland Red West
F472RBF	PMT	F615PWS	City Line	FFS7X	Lowland	G133HNP	Midland Red West
F473RBF	PMT	F615RTC	City Line	FFS9X	Lowland	G134HNP	Midland Red West
F475VEH	PMT	F615XWY	Rider Group	FFS10X	Lowland	G135HNP	Midland Red West
F531UVT	PMT	F616PWS	City Line	FJF193	Leicester	G136HNP	Midland Red West
F546EJA	Rider Group	F616RTC	City Line	FNM868Y	King's of Dunblain	G136YRY	PMT
F578OOU	City Line	F616XWY	Rider Group	FPR66V	Rider Group	G137HNP	Midland Red West
F579OOU	City Line	F617PWS	City Line	FRP906T	Badgerline	G137WOW	People's Provincial
F581XWY	Rider Group	F617RTC	City Line	FSU302	Midland Bluebird	G138HNP	Midland Red West
F582XWY	Rider Group	F617XWY	Rider Group	FSU308	Midland Bluebird	G138WOW	People's Provincial
F583XWY	Rider Group	F618RTC	City Line	FSU315	Midland Bluebird	G139HNP	Midland Red West
F584XWY	Rider Group	F618XWY	Rider Group	FSU318	Midland Bluebird	G139WOW	People's Provincial
F585XWY	Rider Group	F619RTC	City Line	FSU333	Mair's	G140HNP	Midland Red West
F586XWY	Rider Group	F619XWY	Rider Group	FSU334	Midland Bluebird	G141HNP	Midland Red West
F587XWY	Rider Group	F620RTC	City Line	FSU335	Mair's	G142HNP	Midland Red West
F588XWY	Rider Group	F620XWY	Rider Group	FSU380	Midland Bluebird	G143HNP	Midland Red West
F589XWY	Rider Group	F621RTC	City Line	FSU381	Midland Bluebird	G143SUS	Mair's
F590OHT	City Line	F621XWY	Rider Group	FSU382	Midland Bluebird	G144HNP	Midland Red West
F590XWY	Rider Group	F622RTC	City Line	FSU383	Midland Bluebird	G145HNP	Midland Red West
F591OHT	City Line	F622XWY	Rider Group	FSV634	Midland Bluebird	G146HNP	Midland Red West
F591XWY	Rider Group	F623RTC	City Line	FUM486Y	Rider Group	G147HNP	Midland Red West
F592OHT	City Line	F624RTC	City Line	FUM487Y	Rider Group	G148HNP	Midland Red West
F592XWY	Rider Group	F625RTC	City Line	FUM489Y	Rider Group	G149HNP	Midland Red West
F593OHT	City Line	F626RTC	City Line	FUM491Y	Rider Group	G150HNP	Midland Red West
F593XWY	Rider Group	F627RTC	City Line	FUM492Y	Rider Group	G151GOL	Western National
F594OHT	City Line	F628RTC	City Line	FUM494Y	Rider Group	G152GOL	Western National
F594XWY	Rider Group	F629RTC	City Line	FUM495Y	Rider Group	G153GOL	Western National
F595OHT	City Line	F630RTC	City Line	FUM498Y	Rider Group	G154GOL	Western National
F595XWY	Rider Group	F631RTC	City Line	FUM499Y	Rider Group	G155GOL	Western National
F596OHT	City Line	F632RTC	City Line	FUT179V	Leicester	G176JYG	Rider Group
F596XWY	Rider Group	F633JSO	Mair's	FUT180V	Leicester	G177JYG	Rider Group
F597OHT	City Line	F634JSO	Mair's	FUT181V	Leicester	G178JYG	Rider Group
F597XWY	Rider Group	F660EBU	PMT	FUT182V	Leicester	G179JYG	Rider Group
F598PWS	City Line	F695AWW	Badgerline	FUT183V	Leicester	G180JYG	Rider Group
F598XWY	Rider Group	F701MBC	Eastern Counties	FUT184V	Leicester	G181JYG	Rider Group

Reg	Operator	Reg	Operator	Reg	Operator	Reg	Operator
G182JYG	Rider Group	G456JYG	Rider Group	GCR727N	People's Provincial	GWR183T	Rider Group
G183JYG	Rider Group	G457JYG	Rider Group	GCR728N	People's Provincial	GWR185T	Rider Group
G184JYG	Rider Group	G458JYG	Rider Group	GFM101X	PMT	GWR186T	Rider Group
G185JYG	Rider Group	G459JYG	Rider Group	GFM102X	PMT	GWR187T	Rider Group
G210KUA	Rider Group	G460JYG	Rider Group	GFM103X	PMT	GWR188T	Rider Group
G211KUA	Rider Group	G477ERF	PMT	GFM104X	PMT	GWR189T	Rider Group
G212KUA	Rider Group	G478ERF	PMT	GFM105X	PMT	GWR190T	Rider Group
G213KUA	Rider Group	G495FFA	PMT	GFM106X	PMT	GWR191T	Rider Group
G214KUA	Rider Group	G532CVT	PMT	GFM108X	PMT	GWU528T	Rider Group
G215KUA	Rider Group	G541LWU	Western National	GFM109X	PMT	H34USO	Grampian
G216KUA	Rider Group	G550ERF	PMT	GHT127	Badgerline	H35USO	Grampian
G217KUA	Rider Group	G601OSH	Lowland	GMB377T	PMT	H36USO	Grampian
G251JYG	Rider Group	G601OWR	Rider Group	GMS297S	Leicester	H37USO	Grampian
G251LWF	Rider Group	G602OWR	Rider Group	GMS302S	Leicester	H38USO	Grampian
G252JYG	Rider Group	G603OWR	Rider Group	GNG708N	Eastern Counties	H39USO	Grampian
G252LWF	Rider Group	G604OWR	Rider Group	GNG709N	Eastern Counties	H68PDW	Wessex
G253JYG	Rider Group	G605OWR	Rider Group	GNG713N	Eastern Counties	H69PDW	Wessex
G253LWF	Rider Group	G606OWR	Rider Group	GNG714N	Eastern Counties	H71PWO	Wessex
G254JYG	Rider Group	G607OWR	Rider Group	GNG715N	Eastern Counties	H72PWO	Wessex
G254LWF	Rider Group	G608OWR	Rider Group	GRA841V	Eastern Counties	H101KVX	Thamesway
G255JYG	Rider Group	G609OWR	Rider Group	GRA842V	Eastern Counties	H102KVX	Thamesway
G255LWF	Rider Group	G610OWR	Rider Group	GRA843V	Eastern Counties	H103KVX	Thamesway
G256LWF	Rider Group	G611OWR	Rider Group	GRA844V	Eastern Counties	H103TSH	Lowland
G257LWF	Rider Group	G612OWR	Rider Group	GRA845V	Eastern Counties	H104KVX	Thamesway
G258LWF	Rider Group	G613OWR	Rider Group	GRA846V	Eastern Counties	H160JRE	PMT
G259LWF	Rider Group	G614OWR	Rider Group	GRA847V	Eastern Counties	H180JRE	PMT
G261LUG	Brewers	G615OWR	Rider Group	GRF267V	Eastern Counties	H189CNS	PMT
G318YVT	PMT	G616OWR	Rider Group	GRF701V	PMT	H193BTC	Wessex
G330XRE	PMT	G617OWR	Rider Group	GRF704V	PMT	H193CVU	Mair's
G331XRE	PMT	G618OWR	Rider Group	GRF706V	PMT	H194BTC	Wessex
G332XRE	PMT	G619OWR	Rider Group	GRF707V	PMT	H195BTC	Wessex
G333XRE	PMT	G620OWR	Rider Group	GRF708V	PMT	H201JHP	Wessex
G334XRE	PMT	G621OWR	Rider Group	GRF709V	PMT	H202JHP	PMT
G335XRE	PMT	G622OWR	Rider Group	GRF710V	PMT	H203JHP	PMT
G336XRE	PMT	G623OWR	Rider Group	GRF711V	PMT	H204JHP	Midland Red West
G337XRE	PMT	G706JAH	Eastern Counties	GRF714V	PMT	H206JHP	Midland Red West
G338XRE	PMT	G707JAH	Eastern Counties	GRF715V	PMT	H207JHP	Midland Red West
G339XRE	PMT	G708JAH	Eastern Counties	GRF716V	PMT	H208JHP	Midland Red West
G340XRE	PMT	G709JAH	Eastern Counties	GRS114E	Grampian	H289VRP	Northampton
G341XRE	PMT	G710JAH	Eastern Counties	GSC852T	SMT	H290VRP	Northampton
G342CBF	PMT	G715OSH	Lowland	GSC854T	Lowland	H291VRP	Northampton
G343CBF	PMT	G727WJU	Leicester	GSC855T	PMT	H292VRP	Northampton
G344CBF	PMT	G728WJU	Leicester	GSC856T	Eastern Counties	H293VRP	Northampton
G344GEP	SWT	G729WJU	Leicester	GSC857T	Eastern Counties	H294VRP	Northampton
G345CBF	PMT	G730WJU	Leicester	GSC859T	SMT	H301LPU	Thamesway
G345GEP	SWT	G731WJU	Leicester	GSC860T	SMT	H302LPU	Thamesway
G346CBF	PMT	G732WJU	Leicester	GSC861T	SMT	H303LPU	Thamesway
G346GEP	SWT	G733WJU	Leicester	GSO80V	Midland Bluebird	H304LPU	Thamesway
G347ERF	PMT	G734WJU	Leicester	GSO81V	Midland Bluebird	H305LPU	Thamesway
G347GEP	SWT	G735WJU	Leicester	GSU338	Midland Bluebird	H306LPU	Thamesway
G348ERF	PMT	G736WJU	Leicester	GSU339	Midland Bluebird	H307LJN	Thamesway
G348JTH	SWT	G737WJU	Leicester	GSU388	Rider Group	H308LJN	Thamesway
G349ERF	PMT	G753XRE	PMT	GSU390	Rider Group	H310LJN	Thamesway
G349JTH	SWT	G754XRE	PMT	GSX868T	Lowland	H311LJN	Thamesway
G350JTH	SWT	G755XRE	PMT	GSX869T	Lowland	H312LJN	Thamesway
G351JTH	SWT	G756XRE	PMT	GSX891T	Lowland	H313LJN	Thamesway
G352JTH	SWT	G757XRE	PMT	GSX892T	Lowland	H314LJN	Thamesway
G353JTH	SWT	G758XRE	PMT	GSX896T	SMT	H315LJN	Thamesway
G354JTH	SWT	G759XRE	PMT	GSX897T	SMT	H317LJN	Thamesway
G355JTH	SWT	G760XRE	PMT	GSX898T	SMT	H319LJN	Thamesway
G356JTH	SWT	G761XRE	PMT	GSX899T	Lowland	H321LJN	Thamesway
G357JTH	SWT	G762XRE	PMT	GSX900T	SMT	H322LJN	Thamesway
G358JTH	SWT	G801JYG	Rider Group	GSX901T	SMT	H324HVT	Western National
G359JTH	SWT	G802JYG	Rider Group	GTA806N	Western National	H324LJN	Thamesway
G360JTH	SWT	G803JYG	Rider Group	GUG533N	Rider Group	H326LJN	Thamesway
G361JTH	SWT	G804JYG	Rider Group	GUG535N	Rider Group	H327LJN	Thamesway
G365JTH	SWT	G805AAD	PMT	GUG542N	Rider Group	H329LJN	Thamesway
G366JTH	SWT	G805JYG	Rider Group	GUG544N	Rider Group	H330LJN	Thamesway
G367MEP	SWT	G841PNW	Brewers	GUG551N	Rider Group	H331LJN	Thamesway
G368MEP	SWT	G860VAY	Western National	GUG554N	Rider Group	H332LJN	Thamesway
G369MEP	SWT	G901TWS	Badgerline	GUG557N	Rider Group	H334LJN	Thamesway
G370MEP	SWT	G902TWS	Badgerline	GUG559N	Rider Group	H335LJN	Thamesway
G371MEP	SWT	G903TWS	Badgerline	GUG560N	Rider Group	H336LJN	Thamesway
G372MEP	SWT	G904TWS	Badgerline	GUG564N	Rider Group	H337LJN	Thamesway
G373MEP	SWT	G905TWS	Badgerline	GUG565N	Rider Group	H338LJN	Thamesway
G388PNV	Wessex	G906TWS	Badgerline	GUG566N	Rider Group	H339LJN	Thamesway
G389PNV	Wessex	G907TWS	Badgerline	GUG567N	Rider Group	H341LJN	Thamesway
G447LKW	Rider Group	G908TWS	Badgerline	GWR176T	Rider Group	H342LJN	Thamesway
G451JYG	Rider Group	G909TWS	Badgerline	GWR177T	Rider Group	H343LJN	Thamesway
G452JYG	Rider Group	G910TWS	Badgerline	GWR178T	Rider Group	H344LJN	Thamesway
G453JYG	Rider Group	G971KTX	Wessex	GWR180T	Rider Group	H345LJN	Thamesway
G454JYG	Rider Group	G972KTX	Wessex	GWR181T	Rider Group	H346LJN	Thamesway
G455JYG	Rider Group	GAG48N	Eastern Counties	GWR182T	Rider Group	H347LJN	Thamesway

Reg	Operator	Reg	Operator	Reg	Operator	Reg	Operator
H348LJN	Thamesway	H396OHK	Thamesway	H620VNW	Rider Group	H810TWX	Rider Group
H349LJN	Thamesway	H397OHK	Thamesway	H621VNW	Rider Group	H825ERV	Western National
H351HRF	PMT	H398OHK	Thamesway	H622VNW	Rider Group	H834GLD	PMT
H351LJN	Thamesway	H461BEU	Wessex	H623VNW	Rider Group	H835GLD	PMT
H352HRF	PMT	H462BEU	Wessex	H624VNW	Rider Group	H836GLD	PMT
H352LJN	Thamesway	H471OSC	Lowland	H625VNW	Rider Group	H838SLS	King's of Dunblain
H353HRF	PMT	H472OSC	Lowland	H627VNW	Rider Group	H841AHS	Rider Group
H353LJN	Thamesway	H473OSC	SMT	H628VNW	Rider Group	H851GRE	PMT
H354HVT	PMT	H474OSC	SMT	H629VNW	Rider Group	H852GRE	PMT
H354LJN	Thamesway	H475OSC	Midland Bluebird	H630VNW	Rider Group	H853GRE	PMT
H355HVT	PMT	H476OSC	Lowland	H631VNW	Rider Group	H854GRE	PMT
H355LJN	Thamesway	H477OSC	SMT	H632VNW	Rider Group	H855GRE	PMT
H356HVT	PMT	H478OSC	SMT	H633VNW	Rider Group	H856GRE	PMT
H356LJN	Thamesway	H479OSC	SMT	H633YHT	City Line	H857GRE	PMT
H357HVT	PMT	H481JRE	PMT	H634VNW	Rider Group	H858GRE	PMT
H357LJN	Thamesway	H481OSC	Midland Bluebird	H634YHT	City Line	H859GRE	PMT
H358JRE	PMT	H482JRE	PMT	H636VNW	Rider Group	H860GRE	PMT
H358LJN	Thamesway	H482OSC	Midland Bluebird	H636YHT	City Line	H861GRE	PMT
H359JRE	PMT	H483JRE	PMT	H637VNW	Rider Group	H891LOX	Western National
H359LJN	Thamesway	H483OSC	Midland Bluebird	H637YHT	City Line	H892LOX	Western National
H361JRE	PMT	H484OSC	Midland Bluebird	H638VNW	Rider Group	H893LOX	Western National
H361LJN	Thamesway	H485OSC	SMT	H638YHT	City Line	H894LOX	Western National
H362JRE	PMT	H486OSC	SMT	H639VNW	Rider Group	H895LOX	Western National
H362LJN	Thamesway	H487OSC	SMT	H639YHT	City Line	H896LOX	Western National
H363JRE	PMT	H488OSC	SMT	H640VNW	Rider Group	H925PMS	Midland Bluebird
H363LJN	Thamesway	H489OSC	SMT	H640YHT	City Line	H926PMS	Midland Bluebird
H364LJN	Thamesway	H490OSC	SMT	H641VNW	Rider Group	H972RSG	Midland Bluebird
H365LJN	Thamesway	H491OSC	SMT	H641YHT	City Line	H973RSG	Midland Bluebird
H366LFA	PMT	H492OSC	SMT	H642VNW	Rider Group	H974RSG	Midland Bluebird
H366LJN	Thamesway	H493OSC	SMT	H642YHT	City Line	H975RSG	Midland Bluebird
H367LFA	PMT	H494OSC	SMT	H643VNW	Rider Group	H976RSG	Midland Bluebird
H367LJN	Thamesway	H495OSC	SMT	H643YHT	City Line	HAH238V	Eastern Counties
H368LFA	PMT	H496OSC	SMT	H644YHT	City Line	HAH239V	Eastern Counties
H368OHK	Thamesway	H497OSC	SMT	H645YHT	City Line	HAH240V	Eastern Counties
H369LFA	PMT	H498OSC	SMT	H646YHT	City Line	HHJ372Y	Eastern Counties
H369OHK	Thamesway	H499OSC	SMT	H647YHT	City Line	HHJ374Y	Brewers
H370LFA	PMT	H501OSC	SMT	H648YHT	City Line	HHJ377Y	Brewers
H370OHK	Thamesway	H502OSC	Lowland	H649USH	Lowland	HHJ379Y	Brewers
H371LFA	PMT	H523CTR	People's Provincial	H649YHT	City Line	HHJ382Y	Eastern National
H371OHK	Thamesway	H601OVW	Eastern National	H650YHT	City Line	HOR413L	People's Provincial
H372MEH	PMT	H602OVW	Eastern National	H651YHT	City Line	HOR414L	People's Provincial
H372OHK	Thamesway	H603OVW	Eastern National	H652YHT	City Line	HOR415L	People's Provincial
H373MVT	PMT	H604OVW	Eastern National	H653YHT	City Line	HOR416L	People's Provincial
H373OHK	Thamesway	H605OVW	Eastern National	H654YHT	City Line	HOR417L	People's Provincial
H374OHK	Thamesway	H606OVW	Eastern National	H655YHT	City Line	HRS261V	Grampian
H374OTH	SWT	H607OVW	Eastern National	H656YHT	City Line	HRS262V	Grampian
H375OHK	Thamesway	H608OVW	Eastern National	H657YHT	City Line	HRS263V	Grampian
H375OTH	SWT	H609OVW	Eastern National	H658YHT	City Line	HRS264V	Grampian
H376OHK	Thamesway	H610YTC	Badgerline	H659YHT	City Line	HRS265V	Grampian
H376OTH	SWT	H611EJF	Leicester	H660YHT	City Line	HRS266V	Grampian
H377OHK	Thamesway	H611RAH	Eastern Counties	H661YHT	City Line	HRS267V	Grampian
H377OTH	SWT	H611VNW	Rider Group	H662YHT	City Line	HRS268V	Grampian
H378OHK	Thamesway	H611YTC	Badgerline	H712HGL	Western National	HRS269V	Grampian
H378OTH	SWT	H612EJF	Leicester	H713HGL	Western National	HRS270V	Grampian
H379OHK	Thamesway	H612RAH	Eastern Counties	H714HGL	Western National	HRS271V	Grampian
H379OTH	SWT	H612VNW	Rider Group	H715HGL	Western National	HRS272V	Grampian
H380OHK	Thamesway	H612YTC	Badgerline	H716HGL	Western National	HRS273V	Grampian
H380OTH	SWT	H613EJF	Leicester	H717HGL	Western National	HRS274V	Grampian
H381OHK	Thamesway	H613RAH	Eastern Counties	H718HGL	Western National	HRS275V	Grampian
H381OTH	SWT	H613UWR	Western National	H719HGL	Western National	HRS276V	Grampian
H382OHK	Thamesway	H613VNW	Rider Group	H720HGL	Western National	HRS277V	Grampian
H382TTH	SWT	H613YTC	Badgerline	H721HGL	Western National	HRS278V	Grampian
H383OHK	Thamesway	H614EJF	Leicester	H722HGL	Western National	HRS279V	Grampian
H384OHK	Thamesway	H614RAH	Eastern Counties	H723HGL	Western National	HRS280V	Grampian
H385OHK	Thamesway	H614UWR	Western National	H724HGL	Western National	HSF76X	SMT
H386OHK	Thamesway	H614VNW	Rider Group	H725HGL	Western National	HSF77X	SMT
H387OHK	Thamesway	H614YTC	Badgerline	H726HGL	Western National	HSF78X	SMT
H388MAR	Thamesway	H615EJF	Leicester	H726VNW	Rider Group	HSF80X	SMT
H388OHK	Thamesway	H615RAH	Eastern Counties	H751ENR	Leicester	HSF81X	SMT
H389MAR	Thamesway	H615UWR	Western National	H752ENR	Leicester	HSF82X	SMT
H389OHK	Thamesway	H615VNW	Rider Group	H801GRE	PMT	HSF83X	SMT
H390MAR	Thamesway	H615YTC	Badgerline	H802GRE	PMT	HSF84X	SMT
H390OHK	Thamesway	H616EJF	Leicester	H803GRE	PMT	HSF85X	SMT
H391MAR	Thamesway	H616RAH	Eastern Counties	H804GRE	PMT	HSF86X	SMT
H391OHK	Thamesway	H616VNW	Rider Group	H805GRE	PMT	HSF87X	SMT
H392MAR	Thamesway	H616YTC	Badgerline	H806GRE	PMT	HSF88X	SMT
H392OHK	Thamesway	H617RAH	Eastern Counties	H806TWX	Rider Group	HSF91X	SMT
H393MAR	Thamesway	H617VNW	Rider Group	H807GRE	PMT	HSF92X	SMT
H393OHK	Thamesway	H618RAH	Eastern Counties	H807TWX	Rider Group	HSF93X	SMT
H394MAR	Thamesway	H618VNW	Rider Group	H808GRE	PMT	HSF94X	SMT
H394OHK	Thamesway	H619RAH	Eastern Counties	H808TWX	Rider Group	HSF95X	SMT
H395MAR	Thamesway	H619VNW	Rider Group	H809GRE	PMT	HSO281V	Grampian
H395OHK	Thamesway	H620RAH	Eastern Counties	H809TWX	Rider Group	HSO282V	Grampian

Reg	Operator	Reg	Operator	Reg	Operator	Reg	Operator
HSO283V	Grampian	J484PVT	PMT	J908SEH	PMT	JUB644V	Rider Group
HSO284V	Grampian	J485PVT	PMT	J909SEH	PMT	JUB645V	Rider Group
HSO285V	Grampian	J486PVT	PMT	J910SEH	PMT	JUB646V	Rider Group
HSO286V	Grampian	J503WSX	SMT	J911SEH	PMT	JUB647V	Rider Group
HSO287V	Grampian	J504WSX	SMT	J912SEH	PMT	JUB649V	Rider Group
HSO288V	Grampian	J505WSX	SMT	J913SEH	PMT	JUM192V	Rider Group
HSO289V	Grampian	J506WSX	SMT	J914SEH	PMT	JUM193V	Rider Group
HSO290V	Grampian	J507WSX	SMT	J915SEH	PMT	JUM194V	Rider Group
HSO61N	Mair's	J508WSX	SMT	J916SEH	PMT	JUM195V	Rider Group
HSU247	Midland Bluebird	J509WSX	SMT	J916WVC	Brewers	JUM196V	Rider Group
HSU273	Midland Bluebird	J510WSX	SMT	J917SEH	PMT	JUM197V	Rider Group
HSU301	Midland Bluebird	J511WSX	SMT	J918SEH	PMT	JUM198V	Rider Group
HSU955	Kirkpatrick	J512WSX	SMT	J920HGD	PMT	JUM199V	Rider Group
HTC726N	Durbin Coaches	J530FCL	Eastern Counties	J994GCP	Rider Group	JUM200V	Rider Group
HTC727N	Badgerline	J610UTW	Eastern National	JAH241V	Eastern Counties	JUM201V	Rider Group
HTC728N	Western National	J611UTW	Eastern National	JAH242V	Eastern Counties	JUM202V	Rider Group
HUA606Y	Rider Group	J612UTW	Eastern National	JAH243V	Eastern Counties	JUM203V	Rider Group
HUA607Y	Rider Group	J613UTW	Eastern National	JAH244V	Eastern Counties	JUM204V	Rider Group
HVU521V	Durbin Coaches	J614UTW	Eastern National	JBP129P	People's Provincial	JUM205V	Rider Group
HWT29N	Rider Group	J615UTW	Eastern National	JBP130P	People's Provincial	JUM206V	Rider Group
HWT30N	Rider Group	J616UTW	Eastern National	JBP131P	People's Provincial	JUM207V	Rider Group
HWT31N	Rider Group	J617UTW	Eastern National	JBP132P	People's Provincial	JUM208V	Rider Group
HWT33N	Rider Group	J618UTW	Eastern National	JBP133P	People's Provincial	JUM209V	Rider Group
HWT36N	Rider Group	J619UTW	Eastern National	JBW211Y	Durbin Coaches	JUM210V	Rider Group
HWT44N	Rider Group	J620UTW	Eastern National	JCL808V	Eastern Counties	JUM211V	Rider Group
HWT47N	Rider Group	J621BVG	Eastern Counties	JCL809V	Eastern Counties	JUM212V	Rider Group
HWT50N	Rider Group	J621UTW	Eastern National	JFS166X	Lowland	JUM213V	Rider Group
HWT54N	Rider Group	J622BVG	Eastern Counties	JFS979X	SMT	JUM214V	Rider Group
HWT59N	Rider Group	J622UTW	Eastern National	JFS980X	SMT	JUM215V	Rider Group
HWT60N	Rider Group	J623BVG	Eastern Counties	JFS981X	SMT	JUM532V	Rider Group
HXI578	Durbin Coaches	J623UTW	Eastern National	JFS982X	SMT	JUM78V	Rider Group
IIL1828	SWT	J624BVG	Eastern Counties	JFS983X	SMT	JUM79V	Rider Group
IIL1829	SWT	J624UTW	Eastern National	JFS984X	Lowland	JUM82V	Rider Group
J1SWT	SWT	J625BVG	Eastern Counties	JFS985X	Lowland	JUM86V	Rider Group
J2SWT	SWT	J625UTW	Eastern National	JFS986X	Lowland	JUM88V	Rider Group
J3SWT	SWT	J626UTW	Eastern National	JHJ139V	Thamesway	JUP113T	Western National
J4SWT	SWT	J627UTW	Eastern National	JHJ140V	Eastern National	JWE243W	Western National
J5SWT	SWT	J628UTW	Eastern National	JHJ141V	Thamesway	JWE244W	Brewers
J8SMT	SMT	J629UTW	Eastern National	JHJ142V	Eastern National	JWE245W	Brewers
J11AFC	Mair's	J630UTW	Eastern National	JHJ144V	Thamesway	JWE246W	Brewers
J11GRT	Grampian	J701CWT	Western National	JHJ145V	Thamesway	JWE247W	Brewers
J45SNY	Thamesway	J703CWT	Western National	JHJ146V	Thamesway	JWE248W	Brewers
J46SNY	Thamesway	J727KBC	Western National	JHJ147V	Eastern National	JWE249W	Brewers
J48SNY	Thamesway	J753MFP	Leicester	JHJ150V	Eastern National	JWE250W	Brewers
J54SNY	Thamesway	J754MFP	Leicester	JHU900X	City Line	JWE251W	Brewers
J115MRP	Northampton	J755MFP	Leicester	JHU901X	City Line	JWT758V	Rider Group
J140KPX	People's Provincial	J756MFP	Leicester	JHU902X	City Line	JWT760V	Eastern National
J141KPX	People's Provincial	J757MFP	Leicester	JHU903X	City Line	JWT762V	Rider Group
J142KPX	People's Provincial	J758NNR	Leicester	JHU904X	City Line	K1GRT	Grampian
J143KPX	People's Provincial	J759NNR	Leicester	JHU905X	Badgerline	K1YRL	Rider Group
J144KPX	People's Provincial	J774WLS	Midland Bluebird	JHU906X	Badgerline	K3GRT	Grampian
J145KPX	People's Provincial	J775WLS	Midland Bluebird	JHU907X	Badgerline	K4GRT	Grampian
J146KPX	People's Provincial	J776WLS	Midland Bluebird	JHU908X	Badgerline	K9BMS	Brewers
J203HWS	Wessex	J778WLS	Midland Bluebird	JHU909X	Badgerline	K10BMS	Brewers
J204HWS	Wessex	J779WLS	Midland Bluebird	JHU910X	Badgerline	K11BMS	Brewers
J210GNV	Northampton	J795KHD	SMT	JHU911X	Badgerline	K12BMS	Brewers
J243LGL	Western National	J812KHD	SMT	JHU913X	Badgerline	K13BMS	Brewers
J244LGL	Western National	J850FTC	Badgerline	JHU914X	City Line	K26HCL	Eastern Counties
J245LGL	Western National	J850OBV	Badgerline	JHW107P	Badgerline	K27HCL	Eastern Counties
J246LGL	Western National	J851FTC	Badgerline	JHW108P	Badgerline	K28HCL	Eastern Counties
J295GNV	Northampton	J852FTC	Badgerline	JHW109P	Badgerline	K29HCL	Eastern Counties
J296GNV	Northampton	J853FTC	Badgerline	JHW114P	Badgerline	K29OEU	Badgerline
J297GNV	Northampton	J854FTC	Badgerline	JKW215W	Rider Group	K65OHT	Wessex
J298GNV	Northampton	J855FTC	Badgerline	JKW216W	Rider Group	K67HSA	Mair's
J299GNV	Northampton	J856FTC	Badgerline	JNG49N	Eastern Counties	K67OHT	Wessex
J301ASH	Lowland	J857FTC	Badgerline	JNG50N	Eastern Counties	K101HUM	Rider Group
J302ASH	Lowland	J858FTC	Badgerline	JNG51N	Eastern Counties	K102HUM	Rider Group
J303ASH	Lowland	J859FTC	Badgerline	JNG52N	Eastern Counties	K103HUM	Rider Group
J304ASH	Lowland	J860HWS	Badgerline	JNG54N	Eastern Counties	K104HUM	Rider Group
J310XLS	King's of Dunblain	J861HWS	Badgerline	JNG56N	Eastern Counties	K105HUM	Rider Group
J328RVT	PMT	J862HWS	Badgerline	JNG57N	Eastern Counties	K106HUM	Rider Group
J375WWK	Brewers	J863HWS	Badgerline	JNG58N	Eastern Counties	K107HUM	Rider Group
J421NCP	Rider Group	J864HWS	Badgerline	JNU136N	Badgerline	K108HUM	Rider Group
J422NCP	Rider Group	J865HWS	Badgerline	JNU137N	Eastern Counties	K109HUM	Rider Group
J423NCP	Rider Group	J901MAF	Western National	JNU138N	Western National	K110HUM	Rider Group
J424NCP	Rider Group	J901SEH	PMT	JNW73V	Rider Group	K112HUM	Rider Group
J425NCP	Rider Group	J902SEH	PMT	JSF909T	Midland Bluebird	K113HUM	Rider Group
J426NCP	Rider Group	J903SEH	PMT	JSF922T	SMT	K114HUM	Rider Group
J429GHT	Wessex	J904SEH	PMT	JSF928T	Lowland	K115HUM	Rider Group
J430GHT	Wessex	J905SEH	PMT	JSF929T	Lowland	K116HUM	Rider Group
J430WFA	PMT	J906SEH	PMT	JSV426	Mair's	K117HUM	Rider Group
J431GHT	Wessex	J907SEH	PMT	JTH44W	Western National	K118HUM	Rider Group
J431WFA	PMT			JUB643V	Rider Group	K119HUM	Rider Group

Reg	Operator	Reg	Operator	Reg	Operator	Reg	Operator
K120HUM	Rider Group	K443XRF	PMT	K619ORL	Western National	K793OTC	Wessex
K121URP	Northampton	K445XRF	PMT	K619SBC	Leicester	K794OTC	Wessex
K122URP	Northampton	K446XRF	PMT	K620HUG	Rider Group	K801ORL	Western National
K123URP	Northampton	K447XRF	PMT	K620LAE	City Line	K802ORL	Western National
K124URP	Northampton	K448XRF	PMT	K620ORL	Western National	K803ORL	Western National
K125URP	Northampton	K449XRF	PMT	K620SBC	Leicester	K804ORL	Western National
K126URP	Northampton	K461PNR	Wessex	K621HUG	Rider Group	K805DJN	Thamesway
K127GNH	Northampton	K473EDT	Midland Bluebird	K621LAE	City Line	K806DJN	Thamesway
K128GNH	Northampton	K487CVT	PMT	K621ORL	Western National	K807DJN	Thamesway
K129GNH	Northampton	K488CVT	PMT	K621SBC	Leicester	K808DJN	Thamesway
K130GNH	Northampton	K489CVT	PMT	K622HUG	Rider Group	K809DJN	Thamesway
K131GNH	Northampton	K490CVT	PMT	K622LAE	City Line	K810DJN	Thamesway
K132GNH	Northampton	K491CVT	PMT	K622ORL	Western National	K811DJN	Thamesway
K160PPO	People's Provincial	K492CVT	PMT	K622SBC	Leicester	K867NEU	Badgerline
K161PPO	People's Provincial	K506RJX	Rider Group	K623HUG	Rider Group	K868NEU	Badgerline
K162PPO	People's Provincial	K507RJX	Rider Group	K623LAE	City Line	K869NEU	Badgerline
K163PPO	People's Provincial	K509NOU	Wessex	K623ORL	Western National	K870NEU	Badgerline
K164PPO	People's Provincial	K513BSX	SMT	K624HUG	Rider Group	K871NEU	Badgerline
K165PPO	People's Provincial	K514BSX	SMT	K624LAE	City Line	K872NEU	Badgerline
K175YUE	Lowland	K515BSX	SMT	K624ORL	Western National	K873NEU	Badgerline
K211HUM	Rider Group	K516BSX	SMT	K625HUG	Rider Group	K874NEU	Badgerline
K249PCV	Western National	K517BSX	SMT	K625LAE	City Line	K875NEU	Badgerline
K250PCV	Western National	K527RJX	Rider Group	K625ORL	Western National	K876NEU	Badgerline
K251PCV	Western National	K528RJX	Rider Group	K626HUG	Rider Group	K901CVW	Thamesway
K331OAF	Western National	K544XRF	PMT	K626LAE	City Line	K902CVW	Thamesway
K332OAF	Western National	K601HUG	Rider Group	K627HUG	Rider Group	K903CVW	Thamesway
K333OAF	Western National	K601LAE	City Line	K627LAE	City Line	K904CVW	Thamesway
K334OAF	Western National	K601ORL	Western National	K628HUG	Rider Group	K905CVW	Thamesway
K335OAF	Western National	K602HUG	Rider Group	K628LAE	City Line	K906CVW	Thamesway
K336OAF	Western National	K602LAE	City Line	K629HUG	Rider Group	K907CVW	Thamesway
K337OAF	Western National	K602ORL	Western National	K629LAE	City Line	K908CVW	Thamesway
K338OAF	Western National	K603HUG	Rider Group	K630HUG	Rider Group	K909CVW	Thamesway
K339OAF	Western National	K603LAE	City Line	K630LAE	City Line	K910CVW	Thamesway
K340OAF	Western National	K603ORL	Western National	K631GVX	Eastern National	K911CVW	Thamesway
K341OAF	Western National	K604HUG	Rider Group	K631HUG	Rider Group	K912CVW	Thamesway
K342OAF	Western National	K604LAE	City Line	K632GVX	Eastern National	K913CVW	Thamesway
K343OAF	Western National	K604ORL	Western National	K632HUG	Rider Group	K914CVW	Thamesway
K344ORL	Western National	K605HUG	Rider Group	K633GVX	Eastern National	K915CVW	Thamesway
K345ORL	Western National	K605LAE	City Line	K633HUG	Rider Group	K916CVW	Thamesway
K346ORL	Western National	K605ORL	Western National	K634GVX	Eastern National	K917CVW	Thamesway
K347ORL	Western National	K606HUG	Rider Group	K634HUG	Rider Group	K919XRF	PMT
K348ORL	Western National	K606LAE	City Line	K635GVX	Eastern National	K920XRF	PMT
K349ORL	Western National	K606ORL	Western National	K636GVX	Eastern National	K921XRF	PMT
K350ORL	Western National	K607HUG	Rider Group	K637GVX	Eastern National	K922XRF	PMT
K351ORL	Western National	K607LAE	City Line	K638GVX	Eastern National	K923XRF	PMT
K352ORL	Western National	K607ORL	Western National	K639GVX	Eastern National	K924RGE	Midland Bluebird
K353ORL	Western National	K608HUG	Rider Group	K640GVX	Eastern National	K924XRF	PMT
K354ORL	Western National	K608LAE	City Line	K641GVX	Eastern National	K925XRF	PMT
K374BRE	PMT	K608ORL	Western National	K642GVX	Eastern National	K926XRF	PMT
K375BRE	PMT	K609HUG	Rider Group	K643GVX	Eastern National	K927XRF	PMT
K396KHJ	Thamesway	K609LAE	City Line	K644GVX	Eastern National	K928XRF	PMT
K397KHJ	Thamesway	K609ORL	Western National	K645GVX	Eastern National	K929XRF	PMT
K398KHJ	Thamesway	K610HUG	Rider Group	K646GVX	Eastern National	K950HSA	Mair's
K401BAX	Brewers	K610LAE	City Line	K651HUG	Midland Bluebird	K960EU	Wessex
K401HRS	Grampian	K610ORL	Western National	K652DLS	Midland Bluebird	KBZ3627	Lowland
K402BAX	Brewers	K611HUG	Rider Group	K653DLS	Midland Bluebird	KBZ3628	Lowland
K402HRS	Grampian	K611LAE	City Line	K654DLS	Midland Bluebird	KBZ3629	Lowland
K403BAX	Brewers	K611ORL	Western National	K655DLS	Midland Bluebird	KEP829X	Eastern Counties
K403HRS	Grampian	K612HUG	Rider Group	K656DLS	Midland Bluebird	KEX532	Lowland
K404BAX	Brewers	K612LAE	City Line	K657DLS	Midland Bluebird	KFM111Y	PMT
K404HRS	Grampian	K612ORL	Western National	K658DLS	Midland Bluebird	KFM112Y	PMT
K405BAX	Brewers	K613HUG	Rider Group	K659DLS	Midland Bluebird	KFM113Y	PMT
K405HRS	Grampian	K613LAE	City Line	K731JAH	Eastern Counties	KFM114Y	PMT
K406BAX	Brewers	K613ORL	Western National	K732JAH	Eastern Counties	KFM115Y	PMT
K406HRS	Grampian	K614HUG	Rider Group	K733JAH	Eastern Counties	KJD511P	People's Provincial
K407BAX	Brewers	K614LAE	City Line	K734JAH	Eastern Counties	KJD528P	People's Provincial
K407HRS	Grampian	K614ORL	Western National	K735JAH	Eastern Counties	KKE731N	Eastern Counties
K408BAX	Brewers	K615HUG	Rider Group	K736JAH	Eastern Counties	KKE732N	Eastern Counties
K408HRS	Grampian	K615LAE	City Line	K737JAH	Eastern Counties	KKE733N	Eastern Counties
K409BAX	Brewers	K615ORL	Western National	K738JAH	Eastern Counties	KKE734N	Eastern Counties
K409HRS	Grampian	K616HUG	Rider Group	K739JAH	Eastern Counties	KOO785V	Western National
K410BAX	Brewers	K616LAE	City Line	K740JAH	Eastern Counties	KOO787V	Eastern National
K432XRF	PMT	K616ORL	Western National	K741JAH	Eastern Counties	KOO789V	Eastern National
K433XRF	PMT	K617HUG	Rider Group	K742JAH	Eastern Counties	KOO790V	Eastern National
K434XRF	PMT	K617LAE	City Line	K743JAH	Eastern Counties	KOO791V	Badgerline
K435XRF	PMT	K617ORL	Western National	K744JAH	Eastern Counties	KOO792V	Badgerline
K436XRF	PMT	K617SBC	Leicester	K746VJU	Leicester	KOO793V	Badgerline
K437XRF	PMT	K618HUG	Rider Group	K748VJU	Leicester	KOO794V	Eastern National
K438XRF	PMT	K618LAE	City Line	K749VJU	Leicester	KOU791P	Badgerline
K439XRF	PMT	K618ORL	Western National	K750VJU	Leicester	KOU792P	Badgerline
K440XRF	PMT	K618SBC	Leicester	K760SBC	Leicester	KOU794P	Badgerline
K441XRF	PMT	K619HUG	Rider Group	K761SBC	Leicester	KOU795P	Badgerline
K442XRF	PMT	K619LAE	City Line	K792OTC	Wessex	KPJ253W	Rider Group

Reg.	Operator	Reg.	Operator	Reg.	Operator	Reg.	Operator
KPJ255W	Rider Group	L123PWR	Rider Group	L218AAB	Midland Red West	L455LVT	PMT
KPJ257W	Rider Group	L123TFB	Badgerline	L218VHU	Badgerline	L493HRE	PMT
KPJ260W	Rider Group	L124PWR	Rider Group	L219AAB	Midland Red West	L494HRE	PMT
KPJ261W	Rider Group	L124TFB	Badgerline	L219VHU	Badgerline	L495HRE	PMT
KPJ263W	Rider Group	L125PWR	Rider Group	L220AAB	Midland Red West	L496HRE	PMT
KPJ290W	Rider Group	L125TFB	Badgerline	L220VHU	Badgerline	L497HRE	PMT
KPJ291W	Rider Group	L126PWR	Rider Group	L221AAB	Midland Red West	L498HRE	PMT
KPJ292W	Rider Group	L126TFB	Badgerline	L221VHU	Badgerline	L501HCY	SWT
KSA192P	SMT	L127PWR	Rider Group	L223AAB	Midland Red West	L501KSA	Grampian
KSU388	Lowland	L127TFB	Badgerline	L223VHU	Badgerline	L501VHU	City Line
KSU389	Lowland	L128PWR	Rider Group	L224AAB	Midland Red West	L502HCY	SWT
KSU390	Lowland	L128TFB	Badgerline	L224VHU	Badgerline	L502KSA	Grampian
KSU391	Lowland	L129PWR	Rider Group	L225AAB	Midland Red West	L502VHU	City Line
KSU392	Lowland	L129TFB	Badgerline	L225VHU	Badgerline	L503HCY	SWT
KSU393	Lowland	L130PWR	Rider Group	L226AAB	Midland Red West	L503KSA	Grampian
KSU394	Lowland	L130TFB	Badgerline	L227AAB	Midland Red West	L503VHU	City Line
KSU834	Midland Bluebird	L131TFB	Badgerline	L228AAB	Midland Red West	L504HCY	SWT
KTT808P	Western National	L132TFB	Badgerline	L229AAB	Midland Red West	L504KSA	Grampian
KUB548V	Rider Group	L133TFB	Badgerline	L230AAB	Midland Red West	L504VHU	City Line
KUX233W	Lowland	L134TFB	Badgerline	L231AAB	Midland Red West	L505HCY	SWT
KVG601V	Eastern Counties	L135TFB	Badgerline	L231NRE	PMT	L505KSA	Grampian
KVG602V	Eastern Counties	L136TFB	Badgerline	L232AAB	Midland Red West	L505VHU	City Line
KVG603V	Eastern Counties	L140MAK	Midland Bluebird	L233AAB	Midland Red West	L506GEP	SWT
KVG604V	Eastern Counties	L166TRV	People's Provincial	L234AAB	Midland Red West	L506HCY	SWT
KVG606V	Eastern Counties	L167TRV	People's Provincial	L235AAB	Midland Red West	L506KSA	Grampian
KVG607V	Eastern Counties	L168TRV	People's Provincial	L236AAB	Midland Red West	L506VHU	City Line
KVG608V	Eastern Counties	L169TRV	People's Provincial	L237AAB	Midland Red West	L507HCY	SWT
KVG609V	Eastern Counties	L170TRV	People's Provincial	L245PAH	Eastern Counties	L507KSA	Grampian
KWY216V	Rider Group	L171TRV	People's Provincial	L246PAH	Eastern Counties	L507VHU	City Line
KWY217V	Rider Group	L172TRV	People's Provincial	L247PAH	Eastern Counties	L508HCY	SWT
KWY218V	Rider Group	L173TRV	People's Provincial	L248PAH	Eastern Counties	L508KSA	Grampian
KWY219V	Rider Group	L174TRV	People's Provincial	L249PAH	Eastern Counties	L508VHU	City Line
KWY220V	Rider Group	L175TRV	People's Provincial	L250PAH	Eastern Counties	L509HCY	SWT
KWY221V	Rider Group	L176TRV	People's Provincial	L251PAH	Eastern Counties	L509KSA	Grampian
KWY222V	Rider Group	L177TRV	People's Provincial	L252PAH	Eastern Counties	L510HCY	SWT
KWY223V	Rider Group	L178TRV	People's Provincial	L253PAH	Eastern Counties	L510KSA	Grampian
KWY224V	Rider Group	L201AAB	Midland Red West	L254PAH	Eastern Counties	L511HCY	SWT
KWY225V	Rider Group	L201KFS	SMT	L254UCV	Western National	L511KSA	Grampian
KWY226V	Rider Group	L201SHW	Badgerline	L255PAH	Eastern Counties	L511NYG	Rider Group
KWY227V	Rider Group	L202AAB	Midland Red West	L255UCV	Western National	L512HCY	SWT
KWY228V	Rider Group	L202KFS	SMT	L256PAH	Eastern Counties	L512KSA	Grampian
KWY229V	Rider Group	L202SHW	Badgerline	L256UCV	Western National	L513HCY	SWT
KWY230V	Rider Group	L203AAB	Midland Red West	L257PAH	Eastern Counties	L513KSA	Grampian
KWY231V	Rider Group	L203KSX	SMT	L257UCV	Western National	L514HCY	SWT
KWY232V	Rider Group	L203SHW	Badgerline	L258PAH	Eastern Counties	L514KSA	Grampian
KWY233V	Rider Group	L204AAB	Midland Red West	L259PAH	Eastern Counties	L515HCY	SWT
KWY234V	Rider Group	L204KSX	SMT	L269GBU	PMT	L516EHD	Rider Group
KWY236V	Rider Group	L204SHW	Badgerline	L301PWR	Rider Group	L516HCY	SWT
KWY237V	Rider Group	L205AAB	Midland Red West	L302PWR	Rider Group	L517EHD	Rider Group
KWY238V	Rider Group	L205KSX	SMT	L303PWR	Rider Group	L517HCY	SWT
KWY239V	Rider Group	L205SHW	Badgerline	L304PWR	Rider Group	L518EHD	Rider Group
KWY241V	Rider Group	L206AAB	Midland Red West	L305PWR	Rider Group	L518HCY	SWT
KWY242V	Rider Group	L206KSX	SMT	L306PWR	Rider Group	L518KSA	SMT
KWY243V	Rider Group	L206SHW	Badgerline	L307PWR	Rider Group	L519HCY	SWT
KWY244V	Rider Group	L207AAB	Midland Red West	L308PWR	Rider Group	L519KSX	SMT
KWY245V	Rider Group	L207KSX	SMT	L309PWR	Rider Group	L520HCY	SWT
KWY246V	Rider Group	L207SHW	Badgerline	L310PWR	Rider Group	L520KSX	SMT
KWY247V	Rider Group	L208AAB	Midland Red West	L311PWR	Rider Group	L521HCY	SWT
KWY248V	Rider Group	L208KSX	SMT	L312PWR	Rider Group	L521KSX	SMT
KWY249V	Rider Group	L208SHW	Badgerline	L313PWR	Rider Group	L522HCY	SWT
KWY250V	Rider Group	L209AAB	Midland Red West	L314PWR	Rider Group	L522KSX	SMT
KWY251V	Rider Group	L209KSX	SMT	L315PWR	Rider Group	L523HCY	SWT
L6BMS	Brewers	L209SHW	Badgerline	L321HRE	PMT	L523KSX	Midland Bluebird
L8BMS	Brewers	L210AAB	Midland Red West	L322AAB	Midland Red West	L524HCY	SWT
L14BMS	Brewers	L210KSX	SMT	L323NRF	PMT	L524KSX	SMT
L21AHA	Eastern National	L210VHU	Badgerline	L355VCV	Western National	L525JEP	SWT
L22YRL	Rider Group	L211AAB	Midland Red West	L356VCV	Western National	L525KSX	SMT
L60HMS	Midland Bluebird	L211KSX	SMT	L357VCV	Western National	L526JEP	SWT
L64UOU	Wessex	L211VHU	Badgerline	L358VCV	Western National	L526KSX	SMT
L65UOU	Wessex	L212AAB	Midland Red West	L359VCV	Western National	L527JEP	SWT
L67UOU	Wessex	L212KSX	SMT	L360VCV	Western National	L527KSX	SMT
L101PWR	Rider Group	L212VHU	Badgerline	L390UHU	Badgerline	L528JEP	SWT
L102PWR	Rider Group	L213AAB	Midland Red West	L401PWR	Rider Group	L529JEP	SWT
L103PWR	Rider Group	L213KSX	SMT	L401VCV	Western National	L530JEP	SWT
L104PWR	Rider Group	L213VHU	Badgerline	L402PWR	Rider Group	L531JEP	SWT
L105PWR	Rider Group	L214AAB	Midland Red West	L402VCV	Western National	L532JEP	SWT
L106PWR	Rider Group	L214VHU	Badgerline	L403PWR	Rider Group	L533JEP	SWT
L109OSX	SMT	L215AAB	Midland Red West	L403VCV	Western National	L534JEP	SWT
L110OSX	SMT	L215VHU	Badgerline	L404PWR	Rider Group	L535JEP	SWT
L121PWR	Rider Group	L216AAB	Midland Red West	L404VCV	Western National	L536JEP	SWT
L121TFB	Badgerline	L216VHU	Badgerline	L405PWR	Rider Group	L537JEP	SWT
L122PWR	Rider Group	L217AAB	Midland Red West	L405VCV	Western National	L538JEP	SWT
L122TFB	Badgerline	L217VHU	Badgerline	L406VCV	Western National	L538XUT	Mair's

Reg	Operator	Reg	Operator	Reg	Operator	Reg	Operator
L539JEP	SWT	L639VCV	Western National	L811OPU	Eastern National	L938LRF	PMT
L540JEP	SWT	L640PWR	Rider Group	L811SAE	City Line	L939LRF	PMT
L541JEP	SWT	L640SEU	City Line	L812OPU	Eastern National	L940LRF	PMT
L541XUT	Rider Group	L640VCV	Western National	L812SAE	City Line	L941LRF	PMT
L542JEP	SWT	L641PWR	Rider Group	L813OPU	Eastern National	L942LRF	PMT
L542XUT	Rider Group	L641SEU	City Line	L813SAE	City Line	LAT662	Lowland
L543JEP	SWT	L641VCV	Western National	L814OPU	Eastern National	LBD922V	Badgerline
L544JEP	SWT	L642PWR	Rider Group	L814SAE	City Line	LBZ2303	Durbin Coaches
L545JEP	SWT	L642SEU	City Line	L815OPU	Eastern National	LBZ2305	Durbin Coaches
L546JEP	SWT	L642VCV	Western National	L815SAE	City Line	LBZ2571	Durbin Coaches
L546XUT	Rider Group	L643PWR	Rider Group	L816HCY	SWT	LBZ2955	Durbin Coaches
L547JEP	SWT	L643SEU	City Line	L816OPU	Eastern National	LEU256P	Badgerline
L548JEP	SWT	L643VCV	Western National	L816SAE	City Line	LEU262P	Durbin Coaches
L549JEP	SWT	L644PWR	Rider Group	L817HCY	SWT	LEU263P	Badgerline
L550JEP	SWT	L644SEU	City Line	L817OPU	Eastern National	LEU269P	Badgerline
L551HMS	Midland Bluebird	L644VCV	Western National	L817SAE	City Line	LFJ841W	Western National
L552GMS	Midland Bluebird	L645PWR	Rider Group	L818HCY	SWT	LFJ842W	Western National
L552HMS	Midland Bluebird	L645SEU	City Line	L818OPU	Eastern National	LFJ843W	Western National
L553GMS	Midland Bluebird	L645VCV	Western National	L818SAE	City Line	LFJ844W	Western National
L553HMS	Midland Bluebird	L646PWR	Rider Group	L819HCY	SWT	LFJ845W	Western National
L553LVT	PMT	L646SEU	City Line	L819OPU	Eastern National	LFJ846W	Western National
L554GMS	Midland Bluebird	L646VCV	Western National	L819SAE	City Line	LFJ847W	Western National
L554HMS	Midland Bluebird	L647MEV	Eastern National	L820HCY	SWT	LFJ867W	Western National
L554LVT	PMT	L647PWR	Rider Group	L820OPU	Eastern National	LFJ871W	Western National
L555GMS	Midland Bluebird	L647SEU	City Line	L820SAE	City Line	LFJ872W	Western National
L555HMS	Midland Bluebird	L647VCV	Western National	L821HCY	SWT	LFJ873W	Western National
L556GMS	Midland Bluebird	L648MEV	Eastern National	L821OPU	Eastern National	LFR293X	Western National
L556HMS	Midland Bluebird	L648PWR	Rider Group	L821SAE	City Line	LHT721P	Badgerline
L556LVT	PMT	L648SEU	City Line	L822HCY	SWT	LIL5068	Brewers
L557GMS	Midland Bluebird	L648VCV	Western National	L822OPU	Eastern National	LIL5069	Brewers
L557JLS	Midland Bluebird	L649MEV	Eastern National	L822SAE	City Line	LIL5070	Brewers
L557LVT	PMT	L649PWR	Rider Group	L823HCY	SWT	LIL5071	Brewers
L558JLS	Midland Bluebird	L649SEU	City Line	L823SAE	City Line	LLT345V	Eastern Counties
L558LVT	PMT	L649VCV	Western National	L824HCY	SWT	LMS374W	Midland Bluebird
L559JLS	Midland Bluebird	L650MEV	Eastern National	L824SAE	City Line	LMS376W	Midland Bluebird
L561JLS	Midland Bluebird	L650PWR	Rider Group	L825HCY	SWT	LMS377W	Midland Bluebird
L562JLS	Midland Bluebird	L650SEU	City Line	L825SAE	City Line	LMS378W	Midland Bluebird
L563JLS	Midland Bluebird	L650VCV	Western National	L826SAE	City Line	LMS379W	Midland Bluebird
L564JLS	Midland Bluebird	L651MEV	Eastern National	L827WHY	City Line	LMS381W	Midland Bluebird
L565JLS	Midland Bluebird	L651PWR	Rider Group	L828WHY	City Line	LMS382W	Midland Bluebird
L566JLS	Midland Bluebird	L651SEU	City Line	L829WHY	City Line	LMS384W	Midland Bluebird
L601FKG	Brewers	L651VCV	Western National	L830WHY	City Line	LMS386W	Midland Bluebird
L601MWC	Thamesway	L652MEV	Eastern National	L862HFA	PMT	LOA832X	Midland Red West
L601PWR	Rider Group	L652PWR	Rider Group	L877TFB	Badgerline	LOI6690	Rider Group
L602FKG	Brewers	L652SEU	City Line	L878VHT	Badgerline	LRS291W	Grampian
L602PWR	Rider Group	L653MEV	Eastern National	L879VHT	Badgerline	LRS292W	Grampian
L603FKG	Brewers	L653PWR	Rider Group	L880VHT	Badgerline	LRS293W	Grampian
L603PWR	Rider Group	L653SEU	City Line	L881VHT	Badgerline	LRS294W	Grampian
L604FKG	Brewers	L654MEV	Eastern National	L883VHT	Badgerline	LRS295W	Grampian
L604PWR	Rider Group	L654PWR	Rider Group	L884VHT	Badgerline	LRS296W	Grampian
L605FKG	Brewers	L654SEU	City Line	L885VHT	Badgerline	LRS297W	Grampian
L605PWR	Rider Group	L655MEV	Eastern National	L886VHT	Badgerline	LRS298W	Grampian
L606FKG	Brewers	L655PWR	Rider Group	L887VHT	Badgerline	LRS299W	Grampian
L607FKG	Brewers	L656MEV	Eastern National	L889VHT	Badgerline	LRS300W	Grampian
L608FKG	Brewers	L707LKY	City Line	L890VHT	Badgerline	LSC932T	SMT
L623XFP	Leicester	L720JKS	Lowland	L891VHT	Badgerline	LSC933T	SMT
L624XFP	Leicester	L721JKS	Lowland	L892VHT	Badgerline	LSC936T	Lowland
L625XFP	Leicester	L722JKS	Lowland	L893VHT	Badgerline	LSC937T	Lowland
L626XFP	Leicester	L723JKS	Lowland	L894VHT	Badgerline	LSC938T	SMT
L628VCV	Western National	L724JKS	Lowland	L895VHT	Badgerline	LSC939T	SMT
L629VCV	Western National	L725JKS	Lowland	L896VHT	Badgerline	LSK475	Grampian
L630VCV	Western National	L726JKS	Lowland	L897VHT	Badgerline	LSK476	Grampian
L631SEU	City Line	L727JKS	Lowland	L898VHT	Badgerline	LSK527	Kirkpatrick
L631VCV	Western National	L801MEV	Eastern National	L899VHT	Badgerline	LSK529	Mair's
L632SEU	City Line	L801SAE	City Line	L901VHT	Badgerline	LSK530	Mair's
L632VCV	Western National	L802MEV	Eastern National	L902VHT	Badgerline	LSK546	Kirkpatrick
L633SEU	City Line	L802SAE	City Line	L903VHT	Badgerline	LSK570	Grampian
L633VCV	Western National	L803OPU	Eastern National	L904VHT	Badgerline	LSK571	Mair's
L634SEU	City Line	L803SAE	City Line	L905VHT	Badgerline	LSK572	Mair's
L634VCV	Western National	L804OPU	Eastern National	L906VHT	Badgerline	LSK573	Kirkpatrick
L635SEU	City Line	L804SAE	City Line	L907VHT	Badgerline	LSU717	Mair's
L635VCV	Western National	L805OPU	Eastern National	L908VHT	Badgerline	LSU788	Durbin Coaches
L636PWR	Rider Group	L805SAE	City Line	L909VHT	Badgerline	LSU917	Mair's
L636SEU	City Line	L806OPU	Eastern National	L910VHT	Badgerline	LTP634R	People's Provincial
L636VCV	Western National	L806SAE	City Line	L911VHT	Badgerline	LTP635R	People's Provincial
L637PWR	Rider Group	L807OPU	Eastern National	L930HFA	PMT	LUA317V	Rider Group
L637SEU	City Line	L807SAE	City Line	L931HFA	PMT	LUA318V	Rider Group
L637VCV	Western National	L808OPU	Eastern National	L932HFA	PMT	LUA319V	Rider Group
L638PWR	Rider Group	L808SAE	City Line	L933HFA	PMT	LUA320V	Rider Group
L638SEU	City Line	L809OPU	Eastern National	L934HFA	PMT	LUA321V	Rider Group
L638VCV	Western National	L809SAE	City Line	L935HFA	PMT	LUA322V	Rider Group
L639PWR	Rider Group	L810OPU	Eastern National	L936HFA	PMT	LUA323V	Rider Group
L639SEU	City Line	L810SAE	City Line	L937LRF	PMT	LUA329V	Rider Group

Reg	Operator	Reg	Operator	Reg	Operator	Reg	Operator
LUA716V	Eastern National	M208VWU	Rider Group	M246MRW	Midland Red West	M379SRE	PMT
LUA717V	Eastern National	M208VWW	Rider Group	M246VWU	Rider Group	M379YEX	Eastern Counties
LUA718V	Rider Group	M209VWU	Rider Group	M246VWW	Rider Group	M380SRE	PMT
LUA719V	Rider Group	M209VWW	Rider Group	M247MRW	Midland Red West	M380YEX	Eastern Counties
LUG113P	Rider Group	M210VWU	Rider Group	M247VWU	Rider Group	M381SRE	PMT
LUG116P	Rider Group	M210VWW	Rider Group	M247VWW	Rider Group	M382SRE	PMT
LUG117P	Rider Group	M211VWU	Rider Group	M248MRW	Midland Red West	M383SRE	PMT
LUG118P	Rider Group	M211VWW	Rider Group	M248VWU	Rider Group	M399OMS	SMT
LWS32Y	Badgerline	M212VWU	Rider Group	M248VWW	Rider Group	M401UUB	Rider Group
LWS42Y	City Line	M212VWW	Rider Group	M249MRW	Midland Red West	M401VWW	Rider Group
LWS43Y	City Line	M213VWU	Rider Group	M249VWU	Rider Group	M402UUB	Rider Group
LWS44Y	City Line	M213VWW	Rider Group	M249VWW	Rider Group	M402VWW	Rider Group
LWS45Y	City Line	M214VWU	Rider Group	M250MRW	Midland Red West	M403UUB	Rider Group
LWU469V	Eastern National	M214VWW	Rider Group	M250VWU	Rider Group	M403VWW	Rider Group
LWU471V	People's Provincial	M215VWU	Rider Group	M250VWW	Rider Group	M404UUB	Rider Group
LWU472V	Rider Group	M215VWW	Rider Group	M251MRW	Midland Red West	M404VWW	Rider Group
M1GRT	Grampian	M216VWU	Rider Group	M251VWU	Rider Group	M405UUB	Rider Group
M25YRE	PMT	M216VWW	Rider Group	M251VWW	Rider Group	M405VWW	Rider Group
M26YRE	PMT	M217VWU	Rider Group	M252MRW	Midland Red West	M406VWW	Rider Group
M27YRE	PMT	M217VWW	Rider Group	M252VWU	Rider Group	M407CCV	Western National
M28YRE	PMT	M218VWU	Rider Group	M252VWW	Rider Group	M407VWW	Rider Group
M41FTC	Wessex	M218VWW	Rider Group	M253MRW	Midland Red West	M408CCV	Western National
M92BOU	Wessex	M219VWU	Rider Group	M253VWU	Rider Group	M408VWW	Rider Group
M101ECV	Western National	M219VWW	Rider Group	M253VWW	Rider Group	M409CCV	Western National
M102ECV	Western National	M220VWU	Rider Group	M254MRW	Midland Red West	M409VWW	Rider Group
M103ECV	Western National	M220VWW	Rider Group	M254VWU	Rider Group	M410CCV	Western National
M106PKS	Lowland	M221EAF	Western National	M254VWW	Rider Group	M410VWW	Rider Group
M107NEP	SWT	M221VWU	Rider Group	M255MRW	Midland Red West	M411CCV	Western National
M108NEP	SWT	M221VWW	Rider Group	M255VWU	Rider Group	M411VWW	Rider Group
M109PWN	SWT	M223VWU	Rider Group	M255VWW	Rider Group	M412CCV	Western National
M110PWN	SWT	M223VWW	Rider Group	M256MRW	Midland Red West	M412VWW	Rider Group
M111PWN	SWT	M224VWU	Rider Group	M256VWU	Rider Group	M413CCV	Western National
M137FAE	Badgerline	M224VWW	Rider Group	M256VWW	Rider Group	M413DEV	Wessex
M138FAE	Badgerline	M225VWU	Rider Group	M257VWU	Rider Group	M413VWW	Rider Group
M139FAE	Badgerline	M225VWW	Rider Group	M257VWW	Rider Group	M414CCV	Western National
M140FAE	Badgerline	M226VWU	Rider Group	M258VWU	Rider Group	M414VWW	Rider Group
M141FAE	Badgerline	M226VWW	Rider Group	M258VWW	Rider Group	M415CCV	Western National
M142FAE	Badgerline	M227VWU	Rider Group	M259VWU	Rider Group	M415VWW	Rider Group
M151PKS	Lowland	M227VWW	Rider Group	M259VWW	Rider Group	M416CCV	Western National
M152PKS	Lowland	M228VWU	Rider Group	M260VWU	Rider Group	M416VWW	Rider Group
M166VJN	Eastern National	M228VWW	Rider Group	M260VWW	Rider Group	M417CCV	Western National
M179XTR	People's Provincial	M229VWU	Rider Group	M261VWU	Rider Group	M417VWW	Rider Group
M180XTR	People's Provincial	M229VWW	Rider Group	M261VWW	Rider Group	M418CCV	Western National
M181XTR	People's Provincial	M230VWU	Rider Group	M262VWU	Rider Group	M418VWW	Rider Group
M182XTR	People's Provincial	M230VWW	Rider Group	M262VWW	Rider Group	M419CCV	Western National
M183XTR	People's Provincial	M231VWU	Rider Group	M263VWU	Rider Group	M419VWW	Rider Group
M184XTR	People's Provincial	M231VWW	Rider Group	M263VWW	Rider Group	M420CCV	Western National
M185XTR	People's Provincial	M232VWU	Rider Group	M264VWW	Rider Group	M420VWW	Rider Group
M186XTR	People's Provincial	M232VWW	Rider Group	M265VWW	Rider Group	M421CCV	Western National
M187XTR	People's Provincial	M233VWU	Rider Group	M266VWW	Rider Group	M421VWW	Rider Group
M188XTR	People's Provincial	M233VWW	Rider Group	M267VWW	Rider Group	M422CCV	Western National
M189XTR	People's Provincial	M234VWU	Rider Group	M268VWW	Rider Group	M422VWW	Rider Group
M190XTR	People's Provincial	M234VWW	Rider Group	M284SMS	Midland Bluebird	M423CCV	Western National
M191XTR	People's Provincial	M235VWU	Rider Group	M290FAE	Badgerline	M423VWW	Rider Group
M192XTR	People's Provincial	M235VWW	Rider Group	M291FAE	Badgerline	M424CCV	Western National
M193XTR	People's Provincial	M236VWU	Rider Group	M292FAE	Badgerline	M424VWW	Rider Group
M194XTR	People's Provincial	M236VWW	Rider Group	M293FAE	Badgerline	M425CCV	Western National
M195XTR	People's Provincial	M237VWU	Rider Group	M294FAE	Badgerline	M425VWW	Rider Group
M196XTR	People's Provincial	M237VWW	Rider Group	M295FAE	Badgerline	M426CCV	Western National
M197XTR	People's Provincial	M238MRW	Midland Red West	M296FAE	Badgerline	M426VWW	Rider Group
M198XTR	People's Provincial	M238VWU	Rider Group	M301BRL	Western National	M427VWW	Rider Group
M199XTR	People's Provincial	M238VWW	Rider Group	M302BRL	Western National	M428VWW	Rider Group
M201VWU	Rider Group	M239MRW	Midland Red West	M303BRL	Western National	M429VWW	Rider Group
M201VWW	Rider Group	M239VWU	Rider Group	M360XEX	Eastern Counties	M430VWW	Rider Group
M201XTR	People's Provincial	M239VWW	Rider Group	M361XEX	Eastern Counties	M431VWW	Rider Group
M202VWU	Rider Group	M240MRW	Midland Red West	M362XEX	Eastern Counties	M432VWW	Rider Group
M202VWW	Rider Group	M240VWU	Rider Group	M363XEX	Eastern Counties	M433VWW	Rider Group
M202XTR	People's Provincial	M240VWW	Rider Group	M364XEX	Eastern Counties	M434VWW	Rider Group
M203VWU	Rider Group	M241MRW	Midland Red West	M365XEX	Eastern Counties	M435VWW	Rider Group
M203VWW	Rider Group	M241VWU	Rider Group	M367XEX	Eastern Counties	M436VWW	Rider Group
M203XTR	People's Provincial	M241VWW	Rider Group	M368XEX	Eastern Counties	M437VWW	Rider Group
M204BPO	People's Provincial	M242MRW	Midland Red West	M369XEX	Eastern Counties	M438VWW	Rider Group
M204VWU	Rider Group	M242VWU	Rider Group	M370XEX	Eastern Counties	M439FHW	Wessex
M204VWW	Rider Group	M242VWW	Rider Group	M371XEX	Eastern Counties	M439VWW	Rider Group
M205BPO	People's Provincial	M243MRW	Midland Red West	M372XEX	Eastern Counties	M440FHW	Wessex
M205VWU	Rider Group	M243VWU	Rider Group	M373XEX	Eastern Counties	M440VWW	Rider Group
M205VWW	Rider Group	M243VWW	Rider Group	M374XEX	Eastern Counties	M447VWW	Rider Group
M206BPO	People's Provincial	M244MRW	Midland Red West	M375YEX	Eastern Counties	M448VWW	Rider Group
M206VWU	Rider Group	M244VWU	Rider Group	M376YEX	Eastern Counties	M449VWW	Rider Group
M206VWW	Rider Group	M244VWW	Rider Group	M377SRE	PMT	M450VWW	Rider Group
M207BPO	People's Provincial	M245MRW	Midland Red West	M377YEX	Eastern Counties	M501CCV	Western National
M207VWU	Rider Group	M245VWU	Rider Group	M378SRE	PMT	M501GRY	Leicester
M207VWW	Rider Group	M245VWW	Rider Group	M378YEX	Eastern Counties	M502CCV	Western National

Reg	Operator	Reg	Operator	Reg	Operator	Reg	Operator
M502GRY	Leicester	M660VJN	Eastern National	M936TEV	Thamesway	MUA873P	Badgerline
M503CCV	Western National	M661VJN	Eastern National	M937TEV	Thamesway	MUA874P	Badgerline
M503GRY	Leicester	M662VJN	Eastern National	M938TEV	Thamesway	MUT206W	Leicester
M504GRY	Leicester	M663VJN	Eastern National	M939TEV	Thamesway	MUT226W	Leicester
M505GRY	Leicester	M664VJN	Eastern National	M940TEV	Thamesway	MUT229W	Leicester
M506GRY	Leicester	M665VJN	Eastern National	M941TEV	Thamesway	MUT251W	Leicester
M507GRY	Leicester	M667VJN	Eastern National	M942TEV	Thamesway	MUT252W	Leicester
M508GRY	Leicester	M668VJN	Eastern National	M943SRE	PMT	MUT253W	Leicester
M509DHU	City Line	M669VJN	Eastern National	M943TEV	Thamesway	MUT254W	Leicester
M509GRY	Leicester	M670VJN	Eastern National	M944SRE	PMT	MUT255W	Leicester
M510DHU	City Line	M671VJN	Eastern National	M945SRE	PMT	MUT256W	Leicester
M510GRY	Leicester	M672VJN	Eastern National	M946SRE	PMT	MUT257W	Leicester
M511DHU	City Line	M673VJN	Eastern National	M947SRE	PMT	MUT259W	Leicester
M512DHU	City Line	M674VJN	Eastern National	M948SRE	PMT	MUT260W	Leicester
M513DHU	City Line	M675VJN	Eastern National	M949SRE	PMT	MUT261W	Leicester
M514DHU	City Line	M676VJN	Eastern National	M951SRE	PMT	MUT262W	Leicester
M515DHU	City Line	M763CWS	Wessex	M952SRE	PMT	MUT263W	Leicester
M516DHU	City Line	M764CWS	Wessex	M953XVT	PMT	MUT264W	Leicester
M516RSS	Grampian	M765CWS	Wessex	M954XVT	PMT	N41RRP	Northampton
M517DHU	City Line	M831ATC	City Line	M955XVT	PMT	N42RRP	Northampton
M517RSS	Grampian	M832ATC	City Line	M956XVT	PMT	N43RRP	Northampton
M518DHU	City Line	M833ATC	City Line	M957XVT	PMT	N62CSC	SMT
M518RSS	Grampian	M834ATC	City Line	M958XVT	PMT	N63CSC	SMT
M519DHU	City Line	M835ATC	City Line	M959XVT	PMT	N64CSC	SMT
M519RSS	Grampian	M836ATC	City Line	M960XVT	PMT	N65CSC	SMT
M520FFB	City Line	M837ATC	City Line	M961XVT	PMT	N66CSC	SMT
M520RSS	Grampian	M838ATC	City Line	M962XVT	PMT	N67CSC	SMT
M521FFB	City Line	M839ATC	City Line	M963XVT	PMT	N68CSC	SMT
M521RSS	Grampian	M840ATC	City Line	M964XVT	PMT	N69CSC	SMT
M522FFB	City Line	M841ATC	City Line	M965XVT	PMT	N70CSC	SMT
M522RSS	Grampian	M842ATC	City Line	M966XVT	PMT	N226KAE	Badgerline
M523FFB	City Line	M843ATC	City Line	M967XVT	PMT	N227KAE	Badgerline
M523RSS	Grampian	M844ATC	City Line	M968XVT	PMT	N228KAE	Badgerline
M524FFB	City Line	M845ATC	City Line	M969XVT	PMT	N229KAE	Badgerline
M524RSS	Grampian	M846ATC	City Line	M970XVT	PMT	N230KAE	Badgerline
M525FFB	City Line	M847ATC	City Line	M971XVT	PMT	N231KAE	Badgerline
M526FFB	City Line	M848ATC	City Line	M972XVT	PMT	N232KAE	Badgerline
M527FFB	City Line	M849ATC	City Line	MCL938P	Eastern Counties	N233KAE	Badgerline
M528FFB	City Line	M850ATC	City Line	MCL940P	Eastern Counties	N234KAE	Badgerline
M529FFB	City Line	M851ATC	City Line	MCL941P	Eastern Counties	N235KAE	Badgerline
M530FFB	City Line	M852ATC	City Line	MCL944P	Eastern Counties	N236KAE	Badgerline
M531FFB	City Line	M853ATC	City Line	MDL650R	Lowland	N237KAE	Badgerline
M532FFB	City Line	M854ATC	City Line	MEL551P	Rider Group	N238KAE	Badgerline
M533FFB	City Line	M855ATC	City Line	MEX768P	Eastern Counties	N239KAE	Badgerline
M534FFB	City Line	M856ATC	City Line	MEX769P	Eastern Counties	N240KAE	Badgerline
M535FFB	City Line	M857ATC	City Line	MEX770P	Eastern Counties	N241KAE	Badgerline
M536FFB	City Line	M858ATC	City Line	MFA717V	PMT	N242KAE	Badgerline
M537FFB	City Line	M859ATC	City Line	MFA718V	PMT	N301XAB	Midland Red West
M538FFB	City Line	M860ATC	City Line	MFA720V	PMT	N302XAB	Midland Red West
M559SRE	PMT	M861ATC	City Line	MFA721V	Eastern National	N303XAB	Midland Red West
M561SRE	PMT	M862ATC	City Line	MFA723V	PMT	N304XAB	Midland Red West
M562SRE	PMT	M863ATC	City Line	MHJ723V	Rider Group	N305XAB	Midland Red West
M563SRE	PMT	M864ATC	City Line	MHJ726V	Rider Group	N306XAB	Midland Red West
M564SRE	PMT	M865ATC	City Line	MHJ729V	Eastern National	N307XAB	Midland Red West
M565SRE	PMT	M866ATC	City Line	MHJ729V	Rider Group	N308XAB	Midland Red West
M566SRE	PMT	M867ATC	City Line	MHJ731V	Eastern National	N309XAB	Midland Red West
M567RMS	Midland Bluebird	M868ATC	City Line	MJT880P	People's Provincial	N310XAB	Midland Red West
M567SRE	PMT	M869ATC	City Line	MKH48A	Brewers	N311XAB	Midland Red West
M568RMS	Midland Bluebird	M870ATC	City Line	MKH49A	Brewers	N312XAB	Midland Red West
M568SRE	PMT	M871ATC	City Line	MKH59A	Brewers	N313XAB	Midland Red West
M569RMS	Midland Bluebird	M872ATC	City Line	MKH60A	Brewers	N406ENW	Rider Group
M569SRE	PMT	M873ATC	City Line	MKH69A	Brewers	N407ENW	Rider Group
M570RMS	Midland Bluebird	M874ATC	City Line	MKH87A	Brewers	N408ENW	Rider Group
M570SRE	PMT	M880ATC	City Line	MKH98A	Brewers	N409ENW	Rider Group
M571RMS	Midland Bluebird	M882BEU	Badgerline	MKH487A	SWT	N410ENW	Rider Group
M571SRE	PMT	M918TEV	Thamesway	MKH774A	Brewers	N411ENW	Rider Group
M572SRE	PMT	M919TEV	Thamesway	MKH831A	Brewers	N412ENW	Rider Group
M573SRE	PMT	M920TEV	Thamesway	MKH889A	Brewers	N413ENW	Rider Group
M582DAF	Western National	M921TEV	Thamesway	MNW30P	Rider Group	N414ENW	Rider Group
M584ANG	Eastern Counties	M922TEV	Thamesway	MNW33P	Rider Group	N415ENW	Rider Group
M585ANG	Eastern Counties	M923TEV	Thamesway	MNW38P	Rider Group	N416ENW	Rider Group
M586ANG	Eastern Counties	M924TEV	Thamesway	MNW130V	Rider Group	N417ENW	Rider Group
M587ANG	Eastern Counties	M925TEV	Thamesway	MNW132V	Rider Group	N418ENW	Rider Group
M588ANG	Eastern Counties	M926TEV	Thamesway	MOD571P	People's Provincial	N419ENW	Rider Group
M589ANG	Eastern Counties	M927TEV	Thamesway	MOU747N	City Line	N441ENW	Rider Group
M590ANG	Eastern Counties	M928TEV	Thamesway	MOW636R	People's Provincial	N442ENW	Rider Group
M591ANG	Eastern Counties	M929TEV	Thamesway	MOW637R	People's Provincial	N443ENW	Rider Group
M592ANG	Eastern Counties	M930TEV	Thamesway	MPG293W	Rider Group	N445ENW	Rider Group
M593ANG	Eastern Counties	M931TEV	Thamesway	MTU120Y	PMT	N446ENW	Rider Group
M657VJN	Eastern National	M932TEV	Thamesway	MTU122Y	PMT	N539HAE	City Line
M658VJN	Eastern National	M933TEV	Thamesway	MTU123Y	PMT	N540HAE	City Line
M659VJN	Eastern National	M934TEV	Thamesway	MTU124Y	PMT	N541HAE	City Line
M660SRE	PMT	M935TEV	Thamesway	MTU125Y	PMT	N542HAE	City Line

Reg	Operator	Reg	Operator	Reg	Operator	Reg	Operator
N543HAE	City Line	N614GAH	Eastern Counties	N952CPU	Thamesway	NTC139Y	City Line
N544ENW	Rider Group	N614MHB	Brewers	N953CPU	Thamesway	NTC140Y	City Line
N544HAE	City Line	N615APU	Thamesway	N954CPU	Thamesway	NTC141Y	City Line
N545HAE	City Line	N615GAH	Eastern Counties	N955CPU	Thamesway	NTC142Y	City Line
N546HAE	City Line	N615MHB	Brewers	N956CPU	Thamesway	NTC143Y	City Line
N547HAE	City Line	N616APU	Thamesway	N957CPU	Thamesway	NTC573R	People's Provincial
N548HAE	City Line	N616GAH	Eastern Counties	N958CPU	Thamesway	NTH156X	Western National
N551UCY	SWT	N616MHB	Brewers	N959CPU	Thamesway	NUD801W	Lowland
N552UCY	SWT	N617APU	Thamesway	N960CPU	Thamesway	NUM339V	Rider Group
N553UCY	SWT	N617GAH	Eastern Counties	N961CPU	Thamesway	NWS906R	Wessex
N554UCY	SWT	N617MHB	Brewers	N962CPU	Thamesway	OCK985K	Eastern Counties
N555UCY	SWT	N618APU	Thamesway	N963CPU	Thamesway	OCK988K	Eastern Counties
N556UCY	SWT	N618GAH	Eastern Counties	N964CPU	Thamesway	OCK994K	Eastern Counties
N557UCY	SWT	N618MHB	Brewers	N965CPU	Thamesway	OCK995K	Eastern Counties
N558UCY	SWT	N619APU	Thamesway	N966CPU	Thamesway	ODL657R	Eastern Counties
N559UCY	SWT	N619GAH	Eastern Counties	N967CPU	Thamesway	ODL658R	Eastern Counties
N561UCY	SWT	N61CSC	SMT	N968CPU	Thamesway	ODL659R	Eastern Counties
N562UCY	SWT	N620GAH	Eastern Counties	N969CPU	Thamesway	ODM409V	PMT
N563UCY	SWT	N621GAH	Eastern Counties	N970CPU	Thamesway	ODT232	Western National
N564UCY	SWT	N622GAH	Eastern Counties	N971CPU	Thamesway	OEL233P	Eastern Counties
N565UCY	SWT	N623GAH	Eastern Counties	N972CPU	Thamesway	OEL236P	Eastern Counties
N566UCY	SWT	N624GAH	Eastern Counties	NAH135P	Eastern Counties	OEP791R	SWT
N567UCY	SWT	N625GAH	Eastern Counties	NAH139P	Eastern Counties	OEP792R	SWT
N568UCY	SWT	N626GAH	Eastern Counties	NAH141P	Eastern Counties	OEP794R	Brewers
N572VMS	Midland Bluebird	N627GAH	Eastern Counties	NCK980J	Eastern Counties	OEP795R	Brewers
N573VMS	Midland Bluebird	N628GAH	Eastern Counties	NDL655R	Lowland	OEX792W	Eastern Counties
N574CEH	PMT	N701CPU	Thamesway	NDL656R	Lowland	OEX793W	Eastern Counties
N574VMS	Midland Bluebird	N823APU	Eastern National	NED433W	PMT	OEX794W	Eastern Counties
N575CEH	PMT	N824APU	Eastern National	NEH725W	PMT	OEX796W	Eastern Counties
N576CEH	PMT	N825APU	Eastern National	NEH727W	PMT	OHW489R	Wessex
N577CEH	PMT	N826APU	Eastern National	NEH728W	PMT	OJD195R	Eastern Counties
N578CEH	PMT	N827APU	Eastern National	NEH729W	PMT	OPW181P	Eastern Counties
N579CEH	PMT	N828APU	Eastern National	NEH731W	PMT	ORS201R	SMT
N580CEH	PMT	N829APU	Eastern National	NEH732W	PMT	ORS202R	Midland Bluebird
N581CEH	PMT	N830APU	Eastern National	NFB599R	Wessex	ORS203R	Midland Bluebird
N582CEH	PMT	N851CPU	Thamesway	NFB601R	Wessex	ORS204R	Midland Bluebird
N583CEH	PMT	N852CPU	Thamesway	NFN79M	People's Provincial	ORS205R	Midland Bluebird
N584CEH	PMT	N853CPU	Thamesway	NFP205W	Leicester	ORS206R	Midland Bluebird
N585CEH	PMT	N854CPU	Thamesway	NFX130P	People's Provincial	ORS208R	Midland Bluebird
N586CEH	PMT	N863CEH	PMT	NFX131P	People's Provincial	ORS209R	Midland Bluebird
N587CEH	PMT	N864CEH	PMT	NIB4905	Rider Group	ORS210R	Midland Bluebird
N588CEH	PMT	N865CEH	PMT	NIB4906	Rider Group	ORS211R	Midland Bluebird
N589CEH	PMT	N866CEH	PMT	NIB4908	Rider Group	ORS212R	SMT
N590CEH	PMT	N867CEH	PMT	NLS981W	Midland Bluebird	ORS215R	Midland Bluebird
N591CEH	PMT	N875HWS	City Line	NLS984W	Midland Bluebird	ORS216R	Midland Bluebird
N592CEH	PMT	N876HWS	City Line	NOE539R	Midland Red West	ORS217R	Midland Bluebird
N593CEH	PMT	N877HWS	City Line	NOE541R	Midland Red West	ORS60R	Mair's
N594CEH	PMT	N878HWS	City Line	NOE542R	Midland Red West	OSG51V	Lowland
N601APU	Thamesway	N879HWS	City Line	NOE544R	Midland Red West	OSG52V	Lowland
N601EBP	People's Provincial	N880HWS	City Line	NOE547R	Midland Red West	OSG53V	Lowland
N602APU	Thamesway	N881HWS	City Line	NOE561R	People's Provincial	OSG54V	SMT
N602EBP	People's Provincial	N882HWS	City Line	NOE610R	Midland Red West	OSG55V	Eastern Counties
N603APU	Thamesway	N883HWS	City Line	NPA226W	Rider Group	OSG60V	SMT
N603EBP	People's Provincial	N884HWS	City Line	NPA226W	Rider Group	OSG61V	SMT
N604APU	Thamesway	N885HWS	City Line	NPD146L	People's Provincial	OSG62V	SMT
N604EBP	People's Provincial	N886HWS	City Line	NPD154L	People's Provincial	OSG64V	Lowland
N605APU	Thamesway	N887HWS	City Line	NPU974M	Eastern National	OSG65V	SMT
N605EBP	People's Provincial	N889HWS	City Line	NRS301W	Grampian	OSG66V	SMT
N605GAH	Eastern Counties	N890HWS	City Line	NRS302W	Grampian	OSG69V	SMT
N606APU	Thamesway	N891HWS	City Line	NRS303W	Grampian	OSG71V	Lowland
N606EBP	People's Provincial	N892HWS	City Line	NRS304W	Grampian	OSG72V	Lowland
N606GAH	Eastern Counties	N893HWS	City Line	NRS305W	Grampian	OSG74V	Eastern Counties
N607APU	Thamesway	N894HWS	City Line	NRS306W	Grampian	OSG75V	SMT
N607EBP	People's Provincial	N895HWS	City Line	NRS307W	Grampian	OUP683P	Eastern Counties
N607GAH	Eastern Counties	N896HWS	City Line	NRS308W	Grampian	OVT798	Midland Bluebird
N608APU	Thamesway	N897HWS	City Line	NRS309W	Grampian	OWB243	Western National
N608GAH	Eastern Counties	N898HWS	City Line	NRS310W	Grampian	PCA420V	PMT
N609APU	Thamesway	N899HWS	City Line	NRS311W	Grampian	PCA421V	PMT
N609GAH	Eastern Counties	N901HWS	City Line	NRS312W	Grampian	PCG918M	People's Provincial
N609MHB	Brewers	N902HWS	City Line	NRS313W	Grampian	PCG919M	People's Provincial
N610APU	Thamesway	N903HWS	City Line	NRS314W	Grampian	PCG920M	People's Provincial
N610GAH	Eastern Counties	N904HWS	City Line	NRS315W	Grampian	PCG921M	People's Provincial
N610MHB	Brewers	N905HWS	City Line	NSC411X	Lowland	PCG922M	People's Provincial
N611APU	Thamesway	N906HWS	City Line	NSC413X	Lowland	PCL251W	Eastern Counties
N611GAH	Eastern Counties	N907HWS	City Line	NTC129Y	City Line	PCL252W	Eastern Counties
N611MHB	Brewers	N944CPU	Thamesway	NTC130Y	Badgerline	PCL253W	Eastern Counties
N612APU	Thamesway	N945CPU	Thamesway	NTC131Y	Badgerline	PCL254W	Eastern Counties
N612GAH	Eastern Counties	N946CPU	Thamesway	NTC133Y	City Line	PCL255W	Eastern Counties
N612MHB	Brewers	N947CPU	Thamesway	NTC134Y	City Line	PCL256W	Eastern Counties
N613APU	Thamesway	N948CPU	Thamesway	NTC135Y	City Line	PCL257W	Eastern Counties
N613GAH	Eastern Counties	N949CPU	Thamesway	NTC136Y	City Line	PEU512R	Badgerline
N613MHB	Brewers	N950CPU	Thamesway	NTC137Y	City Line	PEU513R	Badgerline
N614APU	Thamesway	N951CPU	Thamesway	NTC138Y	City Line	PEU514R	Badgerline

Reg	Operator	Reg	Operator	Reg	Operator	Reg	Operator
PEU517R	Badgerline	PUA276W	Rider Group	RHT512S	People's Provincial	SSX614V	SMT
PEU518R	Badgerline	PUA277W	Rider Group	RJI2720	Durbin Coaches	SSX616V	SMT
PEV690R	Wessex	PUA278W	Rider Group	RJI2721	Durbin Coaches	SSX617V	SMT
PEV706R	Rider Group	PUA279W	Rider Group	RJI2723	Durbin Coaches	SSX618V	SMT
PEX610W	Eastern Counties	PUA280W	Rider Group	RJI5704	Durbin Coaches	SSX619V	SMT
PEX613W	Eastern Counties	PUA282W	Rider Group	RJI5706	Durbin Coaches	SSX620V	SMT
PEX614W	Eastern Counties	PUA283W	Rider Group	RJI8029	Brewers	SSX621V	SMT
PEX615W	Eastern Counties	PUA284W	Rider Group	RJI8030	Brewers	SSX622V	SMT
PEX616W	Eastern Counties	PUA285W	Rider Group	RJI8031	Brewers	SSX623V	SMT
PEX617W	Eastern Counties	PUA286W	Rider Group	RJI8032	Brewers	SSX624V	SMT
PHY697S	Western National	PUA287W	Rider Group	RJT147R	People's Provincial	SSX625V	SMT
PJI5625	Durbin Coaches	PUA288W	Rider Group	RJT148R	People's Provincial	SSX627V	SMT
PLS536W	Lowland	PUA289W	Rider Group	RKA869T	Rider Group	SSX628V	SMT
PNW598W	Rider Group	PUA290W	Rider Group	RLG430V	PMT	SSX629V	SMT
PNW601W	Rider Group	PUA291W	Rider Group	RMA443V	PMT	SSX630V	SMT
PNW603W	Rider Group	PUA292W	Rider Group	RMS398W	Midland Bluebird	STW19W	Rider Group
PRC848X	Eastern Counties	PUA293W	Rider Group	RMS399W	Midland Bluebird	STW21W	Eastern National
PRC850X	Eastern Counties	PUA294W	Rider Group	RMS400W	Midland Bluebird	STW22W	Eastern National
PRC851X	Eastern Counties	PUA295W	Rider Group	ROG550Y	Western National	STW23W	Eastern National
PRC852X	Eastern Counties	PUA296W	Rider Group	RPW189R	Eastern Counties	STW26W	Badgerline
PRC853X	Eastern Counties	PUA297W	Rider Group	RRB116R	Eastern Counties	STW27W	Eastern National
PRC854X	Eastern Counties	PUA298W	Rider Group	RSX591V	SMT	STW28W	Eastern National
PRC855X	Eastern Counties	PUA299W	Rider Group	RSX593V	SMT	STW29W	Badgerline
PRC857X	Eastern Counties	PUA300W	Rider Group	RSX594V	SMT	STW31W	Badgerline
PSF311Y	SMT	PUA301W	Rider Group	RSX84J	Lowland	STW32W	Badgerline
PSF312Y	SMT	PUA302W	Rider Group	RTH929S	Western National	STW33W	Badgerline
PSF313Y	Lowland	PUA303W	Rider Group	RTH931S	Badgerline	STW34W	Badgerline
PSF314Y	Lowland	PUA304W	Rider Group	RTH932S	Badgerline	STW36W	Eastern National
PSF315Y	Lowland	PUA305W	Rider Group	RUF37R	People's Provincial	STW37W	Eastern National
PSF316Y	Lowland	PUA306W	Rider Group	RUH346	Western National	STW38W	Eastern National
PSU314	Lowland	PUA307W	Rider Group	RWT534R	Rider Group	SUA127R	Rider Group
PSU315	Lowland	PUA308W	Rider Group	RWU43R	Rider Group	SUA128R	Rider Group
PSU316	Lowland	PUA309W	Rider Group	RWU46R	Rider Group	SUA130R	Rider Group
PSU317	Lowland	PUA310W	Rider Group	RWU54R	Rider Group	SUA132R	Rider Group
PSU318	Lowland	PUA311W	Rider Group	RYG771R	Rider Group	SUA133R	Rider Group
PSU319	Lowland	PUA312W	Rider Group	SAE757S	Wessex	SUA134R	Rider Group
PSU320	Lowland	PUA313W	Rider Group	SBK740S	People's Provincial	SUA140R	Rider Group
PSU321	Lowland	PUA314W	Rider Group	SFA287R	PMT	SUA142R	Rider Group
PSU322	Lowland	PUA316W	Rider Group	SFJ100R	Western National	SUA144R	Rider Group
PSU527	Western National	PUA317W	Rider Group	SFJ101R	People's Provincial	SUA146R	Rider Group
PSU609	Mair's	PUA318W	Rider Group	SFJ105R	Western National	SUA147R	Rider Group
PSU622	Midland Bluebird	PUA319W	Rider Group	SFJ106R	Western National	SUA150R	Rider Group
PSU623	Grampian	PUA320W	Rider Group	SFJ140R	Western National	SUB789W	Eastern National
PSU624	Grampian	PUA321W	Rider Group	SFJ144R	Western National	SWW302R	Eastern National
PSU625	Midland Bluebird	PUA322W	Rider Group	SFJ150R	Western National	SWW305R	Rider Group
PSU626	Mair's	PUA323W	Rider Group	SFJ158R	Western National	SWX533W	Rider Group
PSU627	Mair's	PUA324W	Rider Group	SNG436M	Eastern Counties	SWX535W	Rider Group
PSU628	Mair's	PUA325W	Rider Group	SNG438M	Eastern Counties	SWX537W	Rider Group
PSU629	Mair's	PUA326W	Rider Group	SNG439M	Eastern Counties	SWX538W	Rider Group
PSU630	Kirkpatrick	PUK656R	Midland Red West	SNS830W	Midland Bluebird	SWX540W	Rider Group
PSU631	Mair's	PUM148W	Rider Group	SOA657S	Midland Red West	TAH271W	Eastern Counties
PSU968	Grampian	PUM149W	Rider Group	SOA658S	Midland Red West	TAH272W	Eastern Counties
PTD640S	Rider Group	PVF359R	Eastern Counties	SPR39R	People's Provincial	TAH273W	Eastern Counties
PTD646S	Rider Group	PVF360R	Eastern Counties	SPR40R	People's Provincial	TAH274W	Eastern Counties
PTD649S	Rider Group	PVF367R	Eastern Counties	SPR41R	People's Provincial	TAH275W	Eastern Counties
PTD651S	Rider Group	PVF368R	Eastern Counties	SSC108P	Midland Bluebird	TAH276W	Eastern Counties
PTD658S	Rider Group	PVF369R	Eastern Counties	SSU437	Durbin Coaches	TAH554N	Eastern Counties
PTP239S	People's Provincial	PWY38W	Rider Group	SSU816	Midland Bluebird	TBC40X	Leicester
PTR238S	People's Provincial	PWY41W	Rider Group	SSU821	Midland Bluebird	TBC41X	Leicester
PUA252W	Rider Group	PWY42W	Rider Group	SSU827	Midland Bluebird	TBC42X	Leicester
PUA253W	Rider Group	PWY44W	Eastern National	SSU829	Midland Bluebird	TBC43X	Leicester
PUA254W	Rider Group	PWY587W	Rider Group	SSU831	Midland Bluebird	TBC44X	Leicester
PUA255W	Rider Group	PWY588W	Rider Group	SSU837	Midland Bluebird	TBC45X	Leicester
PUA256W	Rider Group	PXI8935	Midland Bluebird	SSU841	Midland Bluebird	TBC46X	Leicester
PUA257W	Rider Group	Q276UOC	Midland Red West	SSU857	Midland Bluebird	TBC47X	Leicester
PUA258W	Rider Group	Q553UOC	Midland Red West	SSU859	Midland Bluebird	TBC48X	Leicester
PUA259W	Rider Group	RAH129M	Eastern Counties	SSU861	Midland Bluebird	TBC49X	Leicester
PUA260W	Rider Group	RAH134M	Eastern Counties	SSU897	Midland Bluebird	TBC50X	Leicester
PUA261W	Rider Group	RAH258W	Eastern Counties	SSX595V	SMT	TBC51X	Leicester
PUA262W	Rider Group	RAH259W	Eastern Counties	SSX596V	SMT	TBC52X	Leicester
PUA263W	Rider Group	RAH261W	Eastern Counties	SSX597V	SMT	TBC53X	Leicester
PUA264W	Rider Group	RAH262W	Eastern Counties	SSX602V	Lowland	TBC54X	Leicester
PUA265W	Rider Group	RAH263W	Eastern Counties	SSX603V	Lowland	TBC55X	Leicester
PUA266W	Rider Group	RAH266W	Eastern Counties	SSX604V	SMT	TBC56X	Leicester
PUA267W	Rider Group	RAH267W	Eastern Counties	SSX605V	Lowland	TEX401R	Eastern Counties
PUA268W	Rider Group	RAH269W	Eastern Counties	SSX606V	Lowland	TEX402R	Eastern Counties
PUA269W	Rider Group	RAH270W	Eastern Counties	SSX607V	Lowland	TEX403R	Eastern Counties
PUA270W	Rider Group	RFB616S	Badgerline	SSX608V	SMT	TEX404R	Eastern Counties
PUA271W	Rider Group	RFS580V	Midland Bluebird	SSX609V	SMT	TEX406R	Eastern Counties
PUA272W	Rider Group	RFS587V	Midland Bluebird	SSX610V	Lowland	TEX407R	Eastern Counties
PUA273W	Rider Group	RHT503S	People's Provincial	SSX611V	SMT	TEX408R	Eastern Counties
PUA274W	Rider Group	RHT504S	People's Provincial	SSX612V	SMT	THX115S	People's Provincial
PUA275W	Rider Group	RHT511S	Durbin Coaches	SSX613V	SMT	THX131S	People's Provincial

Reg	Operator	Reg	Operator	Reg	Operator	Reg	Operator
THX234S	People's Provincial	ULS338T	Midland Bluebird	VDV121S	Western National	WDM345R	Eastern Counties
THX242S	People's Provincial	ULS620X	Midland Bluebird	VDV137S	Badgerline	WFS145W	Midland Bluebird
THX248S	People's Provincial	ULS622X	Midland Bluebird	VDV141S	Western National	WFS146W	Midland Bluebird
THX531S	Eastern Counties	ULS623X	Midland Bluebird	VDV143S	Badgerline	WFS146W	Midland Bluebird
THX573S	Eastern Counties	ULS624X	Midland Bluebird	VDV144S	Western National	WFS154W	Midland Bluebird
TJN502R	Thamesway	ULS630X	Midland Bluebird	VEX283X	Eastern Counties	WFX253S	People's Provincial
TMS403X	Midland Bluebird	ULS636X	Leicester	VEX284X	Eastern Counties	WFX257S	People's Provincial
TMS408X	Midland Bluebird	ULS637X	Leicester	VEX285X	Eastern Counties	WJN559S	Thamesway
TMS410X	Midland Bluebird	ULS642X	Leicester	VEX286X	Eastern Counties	WJN560S	Thamesway
TMS411X	Midland Bluebird	ULS643X	Midland Bluebird	VEX287X	Eastern Counties	WJN564S	Eastern National
TMS412X	Midland Bluebird	ULS713X	Midland Bluebird	VEX288X	Eastern Counties	WJN565S	Thamesway
TPX41T	People's Provincial	ULS714X	Midland Bluebird	VEX290X	Eastern Counties	WJN566S	Thamesway
TPX42T	People's Provincial	ULS716X	Midland Bluebird	VEX292X	Eastern Counties	WKO131S	Durbin Coaches
TRS333	Grampian	ULS717X	Midland Bluebird	VEX294X	Eastern Counties	WLT724	Midland Bluebird
TSJ593S	SMT	UMB333R	PMT	VEX297X	Eastern Counties	WNC479	Eastern National
TSJ594S	SMT	UNA800S	Rider Group	VEX302X	Eastern Counties	WNC480	Eastern National
TSJ596S	SMT	UOI4323	Rider Group	VFS324V	Lowland	WNO484	SWT
TSJ597S	SMT	UPO443T	People's Provincial	VFX980S	People's Provincial	WNO546L	Eastern National
TSJ598S	SMT	UPO444T	People's Provincial	VFX981S	Eastern Counties	WNW152S	Rider Group
TSJ599S	SMT	URF667S	PMT	VJT738	Western National	WNW156S	Rider Group
TSJ600S	SMT	URF668S	Durbin Coaches	VNO732S	Thamesway	WNW157S	Rider Group
TSJ601S	SMT	URF674S	PMT	VNO745S	Thamesway	WNW158S	Rider Group
TSU651	Grampian	URL94Y	Western National	VOD616S	Western National	WNW159S	Rider Group
TSU682	Midland Bluebird	URL992S	Western National	VOD617S	Western National	WNW161S	Rider Group
TSV612	Midland Bluebird	URS316X	Grampian	VOD618S	Western National	WNW162S	Rider Group
TTC537T	Wessex	URS317X	Grampian	VSU715	Midland Bluebird	WNW163S	Rider Group
TVF620R	Eastern Counties	URS318X	Grampian	VSX756R	Lowland	WNW164S	Rider Group
TWH690T	Rider Group	URS319X	Grampian	VSX758R	Lowland	WNW165S	Rider Group
TWH691T	Rider Group	URS320X	Grampian	VSX759R	Lowland	WNW167S	Rider Group
TWH692T	Rider Group	URS321X	Grampian	VTH941T	Eastern National	WNW169S	Rider Group
TWH693T	Rider Group	URS322X	Grampian	VTH942T	Western National	WNW171S	Rider Group
TWH695T	Rider Group	URS323X	Grampian	VVV70S	Northampton	WNW172S	Rider Group
TWN801S	Brewers	URS324X	Grampian	VWU331X	Rider Group	WNW174S	Rider Group
TWN802S	SWT	URS325X	Grampian	VWU332X	Rider Group	WNW175S	Rider Group
TWN803S	Brewers	URS326X	Grampian	VWW327X	Rider Group	WOC722T	Midland Red West
TWN936S	Western National	URS327X	Grampian	VWW328X	Rider Group	WOC723T	Midland Red West
TWS908T	People's Provincial	URS328X	Grampian	VWW329X	Rider Group	WOC724T	Midland Red West
TWS915T	Western National	URS329X	Grampian	VWW330X	Rider Group	WOV582T	Lowland
TYG742R	Rider Group	URS330X	Grampian	VWW331X	Rider Group	WPW199S	Eastern Counties
TYS255W	Lowland	UTO832S	Western National	VWW332X	Rider Group	WSU447	Mair's
TYS261W	Lowland	UTO836S	People's Provincial	VWW333X	Rider Group	WSU460	Mair's
TYS264W	Lowland	UVF624X	Eastern Counties	VWW334X	Rider Group	WSU479	Midland Bluebird
UAA226M	People's Provincial	UVF625X	Eastern Counties	VWW335X	Rider Group	WSU480	Mair's
UAR586W	Western National	UVF626X	Eastern Counties	VWW336X	Rider Group	WSU481	Mair's
UAR587W	Brewers	UVF627X	Eastern Counties	VWW337X	Rider Group	WSU487	Midland Bluebird
UAR588W	Brewers	UVF628X	Eastern Counties	VWW338X	Rider Group	WSU489	Midland Bluebird
UAR589W	Western National	UVX2S	People's Provincial	VWW339X	Rider Group	WSV135	SMT
UAR590W	Western National	UWA580Y	Durbin Coaches	VWW340X	Rider Group	WSV136	SMT
UAR593W	Eastern National	UWB183	Western National	VWW341X	Rider Group	WSV137	SMT
UAR594W	Western National	UWW18X	Rider Group	VWW342X	Rider Group	WSV138	SMT
UAR595W	Western National	UWW19X	Rider Group	VWW343X	Rider Group	WSV140	SMT
UAR596W	Eastern National	UWW20X	Rider Group	VWW344X	Rider Group	WSV144	SMT
UAR597W	Western National	UWW511X	Rider Group	VWW345X	Rider Group	WSV408	Rider Group
UAR598W	Brewers	UWW520X	Rider Group	VWW346X	Rider Group	WSV409	Rider Group
UAR599W	Eastern National	UWY68X	Rider Group	VWW347X	Rider Group	WSV410	Rider Group
UDM450V	PMT	UWY69X	Rider Group	VWW348X	Rider Group	WTH943T	Western National
UFP175S	Leicester	UWY71X	Rider Group	VWW349X	Rider Group	WTH945T	Western National
UFP233S	Leicester	UWY72X	Rider Group	VWW350X	Rider Group	WTH946T	Western National
UFP239S	Leicester	UWY74X	Rider Group	VWW351X	Rider Group	WTH949T	Eastern National
UFX847S	People's Provincial	UWY75X	Rider Group	VWW352X	Rider Group	WTH950T	Western National
UFX848S	People's Provincial	UWY86X	Rider Group	VWW353X	Rider Group	WTH951T	Western National
UFX854S	Eastern Counties	UWY90X	Rider Group	VWW354X	Rider Group	WTH958T	Eastern National
UFX860S	Badgerline	VAH277X	Eastern Counties	VWW355X	Rider Group	WTH959T	Brewers
UHW661	Western National	VAH281X	Eastern Counties	VWW356X	Rider Group	WTH961T	Western National
ULS100X	SMT	VAH282X	Eastern Counties	VWW357X	Rider Group	WTU465W	PMT
ULS101X	SMT	VAO488Y	Lowland	VWW358X	Rider Group	WTU472W	PMT
ULS102X	SMT	VAR898S	Eastern National	VWW359X	Rider Group	WTU481W	PMT
ULS103X	SMT	VAR899S	Eastern National	VWW360X	Rider Group	WTU482W	PMT
ULS104X	SMT	VAR900S	Thamesway	VWW361X	Rider Group	WTU483W	PMT
ULS105X	SMT	VAY57X	Leicester	VXI8734	Lowland	WTU488W	PMT
ULS106X	SMT	VAY58X	Leicester	VXU444	Midland Bluebird	WTU489W	PMT
ULS107X	SMT	VAY59X	Leicester	WAH586S	Eastern Counties	WTU491W	PMT
ULS108X	SMT	VCA452W	PMT	WAH587S	Eastern Counties	WUM109S	Rider Group
ULS109X	SMT	VCA464W	PMT	WAH588S	Eastern Counties	WUM112S	Rider Group
ULS110X	SMT	VCL461	Brewers	WAH589S	Eastern Counties	WUM120S	Rider Group
ULS111X	SMT	VDH244S	Rider Group	WAH590S	Eastern Counties	WUM121S	Rider Group
ULS112X	SMT	VDV108S	Western National	WAH591S	Eastern Counties	WUM125S	Rider Group
ULS113X	SMT	VDV114S	Western National	WAH592S	Eastern Counties	WUM126S	Rider Group
ULS114X	SMT	VDV116S	Western National	WAH593S	Eastern Counties	WUM443S	Rider Group
ULS115X	SMT	VDV117S	Western National	WAH594S	Eastern Counties	WUM92S	Rider Group
ULS327T	Midland Bluebird	VDV118S	Western National	WAS764V	Midland Bluebird	WUM93S	Rider Group
ULS328T	Leicester	VDV119S	Western National	WCY701	Brewers	WUM97S	Rider Group
ULS332T	Midland Bluebird	VDV120S	Western National	WDM341R	PMT	WVF598S	Eastern Counties

Once upon a time the Bristol VRT was sent south from Scotland in exchange for Bristol FLFs. Three VRs from the Isle of Wight crept north in 1991 and joined the Lowland fleet. Photographed in Edinburgh is 850, MDL650R.
Tony Wilson

WVF599S	Eastern Counties	XJF63Y	Leicester	XSS334Y	Grampian	YR3939	Rider Group
WVT900S	PMT	XJF64Y	Leicester	XSS335Y	Grampian	YRF267X	PMT
WWH26L	Eastern Counties	XJF65Y	Leicester	XSS336Y	Grampian	YRY188T	Leicester
WWN804T	SWT	XJF66Y	Leicester	XSS337Y	Grampian	YRY198T	Leicester
WWN805T	SWT	XJF67Y	Leicester	XSS338Y	Grampian	YRY200T	Leicester
WWN806T	Brewers	XJF68Y	Leicester	XSS339Y	Grampian	YSG631W	Lowland
WWN809T	Brewers	XJF69Y	Leicester	XSS340Y	Grampian	YSG632W	Lowland
WWY118S	Rider Group	XMS421Y	Midland Bluebird	XSS341Y	Grampian	YSG633W	SMT
WWY122S	Rider Group	XMS425Y	Midland Bluebird	XSS342Y	Grampian	YSG634W	SMT
WWY123S	Rider Group	XNG203S	Eastern Counties	XSS343Y	Grampian	YSG635W	SMT
WWY127S	Rider Group	XNG204S	Eastern Counties	XSS344Y	Grampian	YSG636W	SMT
WYY752	Durbin Coaches	XNG205S	Eastern Counties	XSS345Y	Grampian	YSG637W	SMT
XBU15S	Rider Group	XNG206S	Eastern Counties	XWL539	Mair's	YSG638W	SMT
XBU17S	Rider Group	XNG207S	Eastern Counties	XWX181S	Eastern Counties	YSG639W	Lowland
XBU5S	Rider Group	XNG762S	Eastern Counties	YBF685S	PMT	YSG640W	SMT
XBU9S	Rider Group	XNG763S	Eastern Counties	YBK132	Brewers	YSG641W	SMT
XDU178	Leicester	XNG765S	Eastern Counties	YBW487V	PMT	YSG642W	SMT
XDV601S	Western National	XNG766S	Eastern Counties	YBW489V	PMT	YSG643W	SMT
XDV603S	Western National	XNG767S	Eastern Counties	YEV305S	Thamesway	YSG644W	SMT
XDV608S	Western National	XNG768S	Eastern Counties	YEV308S	Eastern National	YSG645W	SMT
XDV609S	Western National	XNG769S	Eastern Counties	YEV309S	Eastern National	YSG646W	SMT
XFF283	Western National	XNG770S	Eastern Counties	YEV311S	Thamesway	YSG647W	SMT
XHK215X	Thamesway	XOV743T	Midland Red West	YEV315S	Thamesway	YSG650W	SMT
XHK217X	Eastern National	XOV744T	Midland Red West	YEV318S	Eastern National	YSG654W	SMT
XHK218X	Thamesway	XOV746T	Midland Red West	YEV319S	Eastern National	YSG655W	SMT
XHK220X	Western National	XOV749T	Midland Red West	YEV320S	Eastern National	YSG656W	SMT
XHK221X	Badgerline	XOV752T	Midland Red West	YEV321S	Eastern National	YSG657W	SMT
XHK222X	Badgerline	XOV758T	Midland Red West	YEV322S	Thamesway	YSG658W	SMT
XHK223X	Western National	XRF1X	PMT	YEV323S	Eastern National	YSG659W	SMT
XHK224X	Badgerline	XRF2X	PMT	YEV325S	Eastern National	YSG660W	SMT
XHK225X	Western National	XSA218S	Midland Bluebird	YEV326S	Thamesway	YSO228T	Grampian
XHK227X	Badgerline	XSA219S	Midland Bluebird	YEV327S	Thamesway	YSO230T	Midland Bluebird
XHK228X	Western National	XSA220S	Midland Bluebird	YEV328S	Eastern National	YSO231T	Grampian
XHK230X	Western National	XSA221S	Midland Bluebird	YEV329S	Thamesway	YSO232T	Mair's
XHK231X	Western National	XSA222S	Midland Bluebird	YFB969V	Wessex	YSO233T	SMT
XHK232X	Eastern National	XSA223S	Midland Bluebird	YFS302W	Midland Bluebird	YSO234T	Grampian
XHK234X	Brewers	XSA224S	Midland Bluebird	YFS303W	Midland Bluebird	YSO235T	Grampian
XHK235X	Thamesway	XSA225S	Midland Bluebird	YFV179R	Eastern Counties	YSO236T	Grampian
XHK236X	Thamesway	XSA226S	SMT	YFY7M	Eastern Counties	YSO237T	Grampian
XHK237X	Thamesway	XSA227S	Midland Bluebird	YHN654M	Eastern Counties	YWW810S	Rider Group
XJF60Y	Leicester	XSS331Y	Grampian	YJF16Y	Midland Bluebird	YWX333X	Rider Group
XJF61Y	Leicester	XSS332Y	Grampian	YJF17T	Kirkpatrick	YYE276T	People's Provincial
XJF62Y	Leicester	XSS333Y	Grampian	YNG211S	Eastern Counties	YYE278T	People's Provincial